Introduction to the
Electronic Properties
of Materials

Introduction to the Electronic Properties of Materials

David Jiles

Ames Laboratory
US Department of Energy

and

Department of Materials Science
and Engineering

and

Department of Electrical
and Computer Engineering
Iowa State University, USA

CHAPMAN & HALL

London · Glasgow · Weinheim · New York · Tokyo · Melbourne · Madras

Published by Chapman & Hall, 2–6 Boundary Row, London SE1 8HN, UK

Chapman & Hall, 2–6 Boundary Row, London SE1 8HN, UK

Blackie Academic & Professional, Wester Cleddens Road, Bishopbriggs, Glasgow G64 2NZ, UK

Chapman & Hall GmbH, Pappelallee 3, 69469 Weinheim, Germany

Chapman & Hall Inc., One Penn Plaza, 41st Floor, New York NY 10119, USA

Chapman & Hall Japan, Thomson Publishing Japan, Hirakawacho Nemoto Building, 6F, 1-7-11 Hirakawa-cho, Chiyoda-ku, Tokyo 102, Japan

Chapman & Hall Australia, Thomas Nelson Australia, 102 Dodds Street, South Melbourne, Victoria 3205, Australia

Chapman & Hall India, R. Seshadri, 32 Second Main Road, CIT East, Madras 600 035, India

First edition 1994

© 1994 David Jiles

Typeset in 10/12pt Times by Thomson Press (India) Ltd, New Delhi
Printed in Great Britain by TJ Press (Padstow) Ltd, Cornwall

ISBN 0 412 49580 5 (HB) 0 412 49590 2 (PB)

A catalogue record for this book is available from the British Library

Library of Congress Cataloging-in-Publication data
Jiles, David.
  Introduction to the electronic properties of materials / David Jiles.
    p.   cm.
  Includes index.
  ISBN 0-412-49580-5. — ISBN 0-412-49590-2
  1. Electronics—Materials. 2. Materials—Electric properties.
I. Title.
TK7871.J55 1993
620.1'1297—dc20
                                         93-35411
                                         CIP

∞ Printed on permanent acid-free text paper, manufactured in accordance with ANSI/NISO Z39.48-1992 and ANSI/NISO Z39.48-1984 (Permanence of Paper).

To Helen, Sarah, Elizabeth and Andrew,
from whom so many hours have been taken...

How soon hath Time, the subtle thief of youth,
stolen on his wing.

Milton

Contents

Preface

The subject of electronics, and in particular the electronic properties of materials, is one which has experienced unprecedented growth in the last thirty years. The discovery of the transistor and the subsequent development of integrated circuits has enabled us to manipulate and control the electronic properties of materials to such an extent that the entire telecommunications and computer industries are dependent on the electronic properties of a few semiconducting materials. The subject area is now so important that no modern physics, materials science or electrical engineering degree programme can be considered complete without a significant lecture course in electronic materials. Ultimately the course requirements of these three groups of students may be quite different, but at the initial stages of the discussion of electronic properties of materials, the course requirements are broadly identical for each of these groups. Furthermore, as the subject continues to grow in importance, the initial teaching of this vital subject needs to occur earlier in the curriculum in order to give the students sufficient time later to cover the increasing amount of material.

It is with these objectives in mind that the present book has been written. It is aimed at undergraduates who have only an introductory knowledge of quantum mechanics. The simplified approach used here enables the subject to be introduced earlier in the curriculum. The goal at each stage has been to present the principles of the behaviour of electrons in materials and to develop a basic understanding with a minimum of technical detail. This has resulted in a discussion in breadth rather than depth, which touches all of the key issues and which provides a secure foundation for further development in more specialized courses at a later stage. The presentation here should be of interest to two groups of students: those who have a primary interest in electronic materials and who need an introductory text as a stepping-stone to more advanced texts; and those whose primary interest lies elsewhere but who would nevertheless benefit from a broad, passing knowledge of the subject.

As with the earlier textbook, *Introduction to Magnetism and Magnetic Materials* (1991) the subject area under discussion here is truly multidisciplinary, spanning the traditional subject areas of physics, electrical engineering and materials science. In writing this book I have striven to keep this in mind in order to maintain the interest of a wider audience. Therefore some of the treatment will seem relatively easy for one group of students while relatively hard for another. Over the entire book however I think that the general mix of subject areas leads to a text that is equally difficult for these three groups of students. Chapters 1–5 could easily be included in a traditional solid-state physics course and should be very familiar to physicists. However chapters 6–10 will appeal more to materials scientists since they will be more familiar with dealing with meso- and macroscopic properties. Finally chapters 11–15 discuss the functional performance of these materials in technological applications which are likely to be of most interest to electrical engineers. These chapters provide a rapid introduction to five important applications of electronic materials, each of which could be further developed in a separate advanced course.

Also, as in *Introduction to Magnetism and Magnetic Materials*, the early chapters of this book contain a number of key exercises for the student to attempt. Completed worked solutions are given at the back of the book. It has been my experience that this is much more useful than simply giving a numerical answer at the back, since if you do not get the problem exactly right under those conditions, you cannot easily find out where you went wrong!

On completion of the text the reader should have gained an understanding of the behaviour of electrons within materials, an appreciation of how the electrons determine the magnetic, thermal, optical and electrical properties of materials and an awareness of how these electronic properties are controlled for use in a number of important technological applications. I hope that the text will provide a useful introduction to more detailed books on the subject and that it will also provide the background for developing the interest of students in this fascinating subject at an early stage in their careers.

Finally, I would like to acknowledge the assistance of several friends and colleagues who have helped me in writing this book. In particular thanks go to M. F. Berard, F. J. Friedlaender, R. D. Greenough, R. L. Gunshor, J. Mallinson, R. W. McCallum, R. E. Newnham, S. B. Palmer and A. H. Silver.

D J
Ames, Iowa

ACKNOWLEDGEMENTS

I am grateful to those publishers credited in captions for permission to reproduce some of the figures in this book.

Glossary of Symbols

A	Area
a	Distance
a	Lattice spacing
α	Mean field constant
	Optical attenuation coefficient
$\boldsymbol{B}$	Magnetic induction (magnetic flux density)
B_{R}	Remanent magnetic induction
B_{s}	Saturation magnetic induction
C	Capacitance
	Curie constant
	Specific heat or heat capacity
C^{e}	Electronic specific heat
C^{l}	Lattice specific heat
C_{v}	Specific heat capacity at constant volume
C_{P}	Specific heat capacity at constant pressure
c	Velocity of light
χ	Magnetic susceptibility
χ_{P}	Pauli paramagnetic susceptibility
$\boldsymbol{D}$	Electric displacement (electric flux density)
$D(\omega)$	Vibrational density of states
	Phonon density of states
$D(E)$	Density of available energy states
d	Diameter
	Distance
δ	Optical penetration depth
E_{F}	Fermi energy
ξ	Electric field strength
E_{B}	Elastic (bulk) modulus
	Binding energy

E_Y	Elastic (Young's) modulus
E_S	Elastic (shear) modulus
E	Energy
E_c	Cohesive energy
e	Electronic charge
	Strain
E_a	Anisotropy energy
E_{ex}	Exchange energy
E_H	Magnetic field energy (Zeeman energy)
ξ_{Hall}	Hall field
E_K	Kinetic energy
E_{loss}	Energy loss
E_P	Potential energy
$E_P(x)$	Potential energy at location x
E_σ	Stress energy
ε	Permittivity (dielectric constant)
ε_1	Real component of dielectric constant (polarization)
ε_2	Imaginary component of dielectric constant (absorption)
ε_0	Permittivity of free space
F_{int}	Internal, or interactive, force
F	Force
F_{app}	Applied force
f	Fermi function
g	Transducer generation coefficient
	Spectroscopic splitting factor
	Lande splitting factor
γ	Gyromagnetic ratio
H	Magnetic field strength
h	Planck's constant
$\hbar$	Planck's constant divided by 2π
H_c	Coercivity
	Critical field
H_{cr}	Remanent coercivity
H_d	Demagnetizing field
H_e	Weiss mean field
H_{eff}	Effective magnetic field
I	Magnetic polarization (intensity of magnetization)
I	Intensity of light
	Electric current

J_Q	Thermal current density
J	Electric current density
	Atomic angular momentum
J	Total atomic angular momentum quantum number
	Exchange constant
j	Total electronic angular momentum quantum number
J_{atom}	Exchange integral for an electron on an atom with electrons on several nearest neighbours
J_{ex}	Exchange integral; exchange interaction between two electrons
K	Anisotropy constant
	Thermal conductivity
k	Optical extinction coefficient
	Interatomic force constant
	Coupling coefficient of transducer
	Wave vector
k_B	Boltzmann's constant
K_{u1}	First anisotropy constant for uniaxial system
K_{u2}	Second anisotropy constant for uniaxial system
K_1	First anisotropy constant for cubic system
K_2	Second anisotropy constant for cubic system
L	Inductance
	Length
	Electronic orbit length
	Macroscopic length of lattice chain
	Length of side of cubic specimen
L	Atomic orbital angular momentum
l	Length
l_0	Unstrained length
Δl	Change in length
l	Orbital angular momentum quantum number
$\mathscr{L}(x)$	Langevin function of x
λ	Wavelength
	Magnetostriction
	Penetration depth in superconductor
λ_d	Penetration depth
λ_t	Transverse magnetostriction
λ_s	Saturation magnetostriction
λ_0	Spontaneous bulk magnetostriction
M	Magnetization
m	Magnetic moment
	Momentum

M	Mass
M_{an}	Anhysteretic magnetization
m_e	Electronic mass
m_l	Orbital magnetic quantum number
m_0	Orbital magnetic moment of electron
M_R	Remanent magnetization
M_0	Saturation magnetization
	(spontaneous magnetization at 0 K)
M_s	Spontaneous magnetization within a domain
m_s	Spin magnetic moment of electron
m_s	Spin magnetic quantum number
m_{tot}	Total magnetic moment of atom
m^*	Effective mass of electrons in bands
μ	Permeability
	Mobility of charge carriers
μ_B	Bohr magneton
μ_0	Permeability of free space
$N(E)$	Density of occupied energy states $(= 2D(E)f(E))$, electron population density
N	Number of atoms per unit volume
	Number of electrons per unit volume
N	Number of turns on coil or solenoid
n	Refractive index
	Principal quantum number
	Number of atoms
N_0	Avogadro's number
$N_0(E)$	Total number of energy states between zero energy and energy E
N^*	Effective number of conduction electrons
v	Frequency $(\omega/2\pi)$
ω	Angular frequency $(2\pi v)$
P	Pressure
P	Polarization
$P(E)$	Probability of occupancy of state with energy E
$P(x)$	Probability of electron being at location x
	Angular momentum operator
P_o	Orbital angular momentum of electron
P_s	Spin angular momentum of electron
P_{tot}	Total angular momentum of electron
p	angular momentum
π	Peltier coefficient
Φ	Magnetic flux
ϕ	Angle
	Work function
	Spin wave function

Ψ	Total wave function
ψ	Electron wave function
q	Electric charge
Q	Quantity of heat
R	Resistance
	Reflectance
$\boldsymbol{r}$	Radius vector
r	Interatomic separation
	Radius
	Radius of ionic cores of atoms in lattice
	Electronic orbit radius
ρ	Density
	Resistivity
ρ_{mag}	Magnetoresistivity
S	Atomic spin angular momentum
S	Entropy
s	Electronic spin angular momentum quantum number
σ	Conductivity
	Stress
T	Temperature
t	Time
	Thickness
T_{c}	Curie temperature
	Critical temperature
t_{o}	Orbital period of electron
θ	Angle
$\boldsymbol{\tau}$	Torque
τ	Time constant
	Relaxation time
τ_{max}	Maximum torque
U	Internal energy
$\boldsymbol{u}$	Unit vector
u	Displacement of an atom from equilibrium
v_{f}	Final velocity (terminal velocity) of electrons in Drude model
v	Velocity
v_{F}	Velocity of electrons at Fermi surface
V	Electrical potential
V	Voltage (potential difference)
	Volume

W	Power
W_a	Atomic weight
W_H	Hysteresis loss
x	Distance along x-axis
y	Distance along y-axis
Z	Impedance
	Atomic number
z	Distance along z-axis
Σ	Number of nearest-neighbour atoms

SI Units, Symbols and Dimensions

Quantity	Unit Symbol	Unit Name	MKSA Base units	Dimensions
Length	m	metre	m	L
Mass	kg	kilogram	kg	M
Time	s	second	s	T
Frequency	Hz	hertz	s^{-1}	T^{-1}
Force	N	newton	$kg\,m\,s^{-2}$	MLT^{-2}
Pressure	Pa	pascal	$kg\,m^{-1}\,s^{-2}$	$ML^{-1}T^{-2}$
Energy	J	joule	$kg\,m^2\,s^{-2}$	ML^2T^{-2}
Power	W	watt	$kg\,m^2\,s^{-3}$	ML^2T^{-3}
Electric charge	C	coulomb	$A\,s$	CT
Electric current	A	ampere	A	C
Electric potential	V	volt	$kg\,m^2\,A^{-1}\,s^{-3}$	$ML^2C^{-1}T^{-3}$
Resistance	Ω	ohm	$kg\,m^2\,A^{-2}\,s^{-3}$	$ML^2C^{-2}T^{-3}$
Resistivity	Ωm	ohm metre	$kg\,m^3\,A^{-2}\,s^{-3}$	$ML^3C^{-2}T^{-3}$
Conductance	S	siemens	$kg^{-1}\,m^{-2}\,A^2\,s^3$	$M^{-1}L^{-2}C^2T^3$
Conductivity	Sm^{-1}	siemens/metre	$A^2\,s^3\,kg^{-1}\,m^{-3}$	$M^{-1}L^{-3}C^2T^3$
Capacitance	F	farad	$A^2\,s^4\,kg^{-1}\,m^{-2}$	$M^{-1}L^{-2}C^2T^4$
Electric displacement (flux density)	$C\,m^{-2}$	coulomb metre^{-2}	$A\,s\,m^{-2}$	$CL^{-2}T$
Electric field	$V\,m^{-1}$	volt metre^{-1}	$kg\,m\,A^{-1}\,s^{-3}$	$MLC^{-1}T^{-3}$
Electric polarization	$C\,m^{-2}$	coulomb metre^{-2}	$A\,s\,m^{-2}$	$CL^{-2}T$
Permittivity	Fm^{-1}	farad/metre	$A^2\,s^4\,kg^{-1}\,m^{-3}$	$M^{-1}L^{-3}C^2T^4$
Inductance	H	henry	$kg\,m^2\,A^{-2}\,s^{-2}$	$ML^2C^{-2}T^{-2}$
Magnetic flux	Wb	weber	$kg\,m^2\,A^{-1}\,s^{-2}$	$ML^2C^{-1}T^{-2}$
Magnetic induction (flux density)	T	tesla	$kg\,A^{-1}\,s^{-2}$	$MC^{-1}T^{-2}$
Magnetic field	$A\,m^{-1}$	ampere/metre	$A\,m^{-1}$	CL^{-1}
Magnetization	$A\,m^{-1}$	ampere metre^{-1}	$A\,m^{-1}$	CL^{-1}
Permeability	$H\,m^{-1}$	henry/metre	$kg\,m\,A^{-2}\,s^{-2}$	$MLC^{-2}T^{-2}$

Values of Selected Physical Constants

Avogadro's number $\qquad$ $N_0 = 6.022 \times 10^{26}$ atoms kgmol^{-1}
Boltzmann's constant $\qquad$ $k_B = 1.381 \times 10^{-23}$ J K^{-1}
Gas constant $\qquad$ $R = 8.314$ J mol^{-1} K^{-1}

Planck's constant $\qquad$ $h = 6.626 \times 10^{-34}$ J s
$\qquad$ $h/2\pi = 1.054 \times 10^{-34}$ J s

Velocity of light in empty
space $\qquad$ $c = 2.998 \times 10^8$ m s^{-1}
Permittivity of empty space $\qquad$ $\varepsilon_0 = 8.854 \times 10^{-12}$ F m^{-1}
Permeability of empty space $\qquad$ $\mu_0 = 1.257 \times 10^{-6}$ H m^{-1}

Atomic mass unit $\qquad$ amu $= 1.661 \times 10^{-27}$ kg

Properties of electrons
$\quad$ Electronic charge $\qquad$ $e = -1.602 \times 10^{-19}$ C
$\quad$ Electronic rest mass $\qquad$ $m_e = 9.109 \times 10^{-31}$ kg
$\quad$ Charge to mass ratio $\qquad$ $e/m_e = 1.759 \times 10^{11}$ C kg^{-1}
$\quad$ Electron volt $\qquad$ eV $= 1.602 \times 10^{-19}$ J

Properties of protons
$\quad$ Proton charge $\qquad$ $e_P = 1.602 \times 10^{-19}$ C
$\quad$ Rest mass $\qquad$ $m_P = 1.673 \times 10^{-27}$ kg
$\quad$ Gyromagnetic ratio of
$\quad$ proton $\qquad$ $\gamma_P = 2.675 \times 10^8$ Hz T^{-1}

Magnetic constants
$\quad$ Bohr magneton $\qquad$ $\mu_B = 9.274 \times 10^{-24}$ A m^2 $(= $ J T$^{-1})$
$\qquad$ $= 1.165 \times 10^{-29}$ Jm A^{-1}
$\quad$ Nuclear magneton $\qquad$ $\mu_N = 5.051 \times 10^{-27}$ A m^2 $(= $ J T$^{-1})$
$\quad$ Magnetic flux quantum $\qquad$ $\Phi_0 = 2.067 \times 10^{-15}$ Wb $(= $ V s$)$

Foreword for the Student

The objective of this book is to present an introduction to the electronic properties of materials that is broad in its coverage but not exhaustive. The book focuses on the understanding of a few basic principles of the behaviour of electrons in materials and uses them to provide a description of a wide range of phenomena including magnetic, electrical, thermal and optical properties of materials. I have also given a number of historical references in the text, particularly in the early chapters. It seems to me that an appreciation of the historical development of a subject helps the overall understanding, apart from which it is interesting to know who originally developed the underlying ideas and even to re-read some of these landmark papers.

It has been my experience that, with the possible exception of the prospective specialist in solid-state physics, the majority of students do not benefit greatly from being confronted with a mass of detailed results arising from the theory of electrons in solids. This can come later for the intending specialist. In introducing this subject it seems more useful to present a few key results based on relatively simple models, which give a general feel for the behaviour of electrons in materials and how they contribute to the observed properties. These models themselves need not be particulary complex to be useful. For example the basic premises of both the classical Drude model and the Sommerfeld model are quite far from reality. Yet the predictions that they make about the properties of the material contain some of the essential known results, for example the Wiedemann–Franz law and the electronic contribution to the heat capacity.

Therefore, the general approach taken here has been to introduce and discuss the consequences of such simple models which can be used to guide our thinking. We begin on the level of a few electrons subjected to an electrostatic potential due to the rest of the material. Subsequently the bulk properties of materials are considered and the phenomena are related to the earlier discussion of the behaviour of electrons. Finally several key applications are discussed, in which the electronic properties of materials play the central role in determining the suitability of materials for these applications. In particular the areas of microelectronics, optoelectronics, superconductivity, magnetism and piezoelectricity are examined.

Part One
Fundamentals of Electrons in Materials

1
Properties of a Material Continuum

The objective of this chapter is simply to remind us of the macroscopic properties of materials and to point out that in uses of electronic materials we are usually exclusively interested in these bulk properties which are the ones which we usually measure. In order to measure these properties we need to give exact definitions of the various quantities. The microscopic properties are mostly only of interest because they help us to explain the variation of the macroscopic properties with external conditions, including any inter-relationships which exist between the macroscopic properties. Once we have achieved an understanding of the relationship between macroscopic properties and the microscopic structure of a material it becomes possible to control the structure in order to produce materials with specific, desired properties.

1.1 RELATIONSHIPS BETWEEN MACROSCOPIC PROPERTIES OF MATERIALS

How do the electrical and thermal properties of a material relate to its optical properties?

If you are shown pieces of two materials, one with a shiny, highly reflecting surface, and the other with a dull surface, which is the good electrical conductor? The shiny one, of course. Which is the good thermal conductor? Again the shiny one. These answers are obvious and are so much ingrained in our everyday experience that they do not even give us much pause for thought. But we should ask ourselves why these answers are correct. After all you did not make any test for electrical or thermal properties of the materials, you only know about their reflectance, an optical property. There must therefore be a very close relationship between the optical, electrical and thermal properties of a material. Furthermore, having decided that the shiny one is a metal based solely on its optical properties, you have probably already decided something about its mechanical properties, since it is quite likely to be more ductile than the material with the dull surface.

These relationships between the various properties of a material must indicate a common mechanism. For example the optical reflectance, electrical conductivity and thermal conductivity of the metal must have a common origin. Yet our macroscopic continuum model of materials gives no indication what this common origin might be. These unexplained relationships between the macroscopic properties of materials form the starting point for our investigation of the electronic properties of materials.

We will begin with some simple definitions of macroscopic properties in order to get our bearings and then consider some of the well-known macroscopic laws obeyed by materials. Our first goal will then be to provide a conceptual framework for understanding these properties and relationships.

1.1.1   Measurable properties of materials

How do we characterize materials in terms of measurable quantities?

On the everyday scale our means of interacting with a solid generally relies on its macroscopic properties, and these as a rule are based on a classical continuum model of the solid. The continuum model makes no assumptions about the internal structure of the material in terms of the nature of its atoms and electrons or their distribution. This model assumes that the material has properties which are uniform throughout, and therefore allows only a single value of such properties as elastic modulus, electrical conductivity, refractive index and thermal conductivity. However, the continuum model does allow the description of anisotropy in properties along certain directions, where appropriate. It is interesting to note that in almost all cases we do not even need to know anything about the internal structure of these materials in order to make use of them. However the behaviour of the various properties of the materials cannot be explained without some understanding of the underlying mechanisms, and this involves the development of microscopic theories of the atomic and electronic structure inside the materials. The relationship between the structure of matter and its physical properties has been treated in detail using only classical physics to describe the materials in the excellent work of Landau, Akhiezer and Lifshitz [1].

The properties of interest depend of course on the application under consideration, but broadly we are usually interested in one or more of the following categories: mechanical, electrical, optical, thermal and magnetic properties. In most cases the material properties are obtained as a result of measurements of two quantities which by themselves do not represent material properties.

Often a measurement is made of the response of a material, in terms of a state parameter (e.g. strain, change in temperature or current density), to the influence of an external effect or field parameter (e.g. stress, amount of heat input or electric field strength). The quotient of these two measurements is then the material property (e.g. elastic modulus, specific heat capacity or electrical conductivity). Compilations of the various macroscopic properties of materials have been made by many authors, of which the most comprehensive is that by Weast [2].

1.1.2 Bulk properties of materials

How can these macroscopic properties be explained?

These bulk, continuum properties of materials are almost exclusively what we are really concerned with in using the materials, because these are the properties which can be directly measured. Although these properties are often documented in great detail for materials, the macroscopic continuum picture gives no explanation of why, for example, copper is a better conductor than glass; why iron is ferromagnetic but manganese is not; why aluminium conducts heat better than sulphur and so on. In order to explain these properties of materials we must look inside the material and try to develop a better understanding of what is happening. These explanations are founded on a description of microscopic rather than macroscopic effects.

1.1.3 Dependence of properties on the environment

Are the material 'constants' really invariant when the external conditions change?

The macroscopic properties of materials such as Young's modulus, thermal and electrical conductivity and magnetic permeability, do not remain constant. The optical parameters k and n are dependent on the wavelength of incident electromagnetic radiation, permeability is dependent on temperature, and so is electrical conductivity.

The elastic modulus of gadolinium, for example, shows unusual behaviour close to 293 K. The variation of the reflectivity of silver with energy of incident electromagnetic radiation reveals a drastic change at about 4 eV. The specific heat of nickel reveals anomalous behaviour at around 600 K, and the magnetic susceptibility of manganese fluoride MnF_2, shows an anomaly in the vicinity of 70 K. All of these show variations in bulk properties which lie beyond explanation on the basis of the continuum theory of matter.

These examples show interesting features in some of the bulk properties of these materials. We will look briefly at a few definitions which are used to quantify the material properties which will concern us and which will be referred to throughout this book.

1.2 MECHANICAL PROPERTIES

How do we quantify the mechanical behaviour of materials?

The mechanical properties broadly encompass the elastic, plastic and acoustic properties of a material. These may be quantified by the following: the bulk modulus E_B, Young's modulus E_Y and the shear modulus E_S. (We use these symbols to avoid possible confusion between the elastic moduli, particularly

Young's modulus which is often given the symbol E, and the energy which we use extensively later, and which also takes the symbol E). The yield strength and the acoustic velocity in different directions, complete the set of mechanical descriptors of a material.

1.2.1   Elastic moduli

How does a material respond to stress?

In a material that is isotropic the elastic properties can be completely specified in terms of two elastic moduli, the longitudinal (or Young's) modulus and the transverse (or shear) modulus. Other elastic properties, such as Poisson's ratio, can be defined in terms of a combination of these two moduli.

Young's modulus, is a material property obtained from measurement of two quantities, the applied longitudinal stress σ and the resulting strain e in the same direction. Since by Hooke's law stress is proportional to strain for small displacements

$$E_Y = \frac{\text{stress}}{\text{strain}} = \frac{\sigma_\parallel}{e_\parallel}.$$

The shear modulus is a material property obtained from measurement of two quantities, the applied shear stress and the resulting shear strain

$$E_S = \frac{\text{shear stress}}{\text{shear strain}} = \frac{\sigma_\perp}{e_\perp}.$$

Table 1.1 gives values of the elastic Young's modulus for various materials. These elastic 'constants' are not actually constant but can vary quite markedly, with temperature for instance. An example of interesting temperature dependence of the elastic modulus in gadolinium is shown in Fig. 1.1. Although the elastic moduli are usually determined principally by the lattice potential between atoms, we will see later that this anomalous behaviour of the elastic modulus of

Table 1.1   Elastic moduli of various materials

Material	*Elastic modulus E_Y (Pa)*	*Shear modulus E_S (Pa)*
Lead	0.16×10^{11}	0.06×10^{11}
Glass	0.55×10^{11}	0.23×10^{11}
Iron	0.91×10^{11}	0.70×10^{11}
Aluminium	0.70×10^{11}	0.24×10^{11}
Nickel	$2.1 \ \times 10^{11}$	0.77×10^{11}
Steel	$2.0 \ \times 10^{11}$	0.84×10^{11}
Tungsten	$3.6 \ \times 10^{11}$	1.50×10^{11}

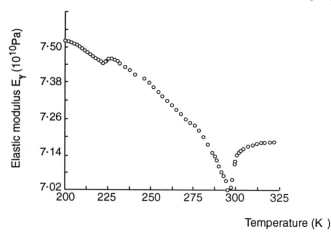

Fig. 1.1 Variation of the elastic modulus of the metal gadolinium with temperature [3].

gadolinium at 293 K and 225 K is the result of reorganization of the electronic magnetic moments during magnetic phase transitions. The electronic interactions perturb the interatomic potential, and at magnetic phase transitions this can cause unusual behaviour of the elastic moduli.

1.3   ELECTRICAL PROPERTIES

How do we quantify the electrical behaviour of materials?

In the case of the electrical properties, we are often concerned with the conductivity. This determines whether we are dealing with an electrical conductor or insulator, for example. In some cases we may be concerned about the electrical polarization as determined by the dielectric constant or permittivity, and in others about the dissipation of electrical energy under AC conditions (eddy currents). The electrical properties of principal interest are: the electrical conductivity σ and the dielectric constant ε.

1.3.1   Ohm's law

Is there a relationship between electric current and electric field strength in a material?

The well known Ohm's law deserves a mention because it provides a test of how well the electronic models of materials perform. It states that the current density

J is related to the electric field ξ in a material by the relationship

$$J = \sigma \xi$$

We use ξ here to distinguish the electric field from energy which is denoted later by E, σ is the electrical conductivity. Alternatively, if the voltage across a material is V volts and the current passing is I amps then,

$$V = IR$$

where R is the resistance of the material [4]. Figure 1.2 shows a range of resistivities for various materials.

1.3.2   Electrical conductivity

How is electric charge transmitted in a material?

The electrical conductivity is the amount of electric charge q transferred per unit time t across unit cross-sectional area A under the action of unit potential gradient dV/dx,

$$\sigma = \frac{(dq/dt)}{A\left(\dfrac{dV}{dx}\right)}$$

and from Ohm's law

$$\sigma = \frac{J}{\xi} = \frac{\text{current density}}{\text{electric field}}.$$

The electrical conductivities of materials exhibit probably the widest range of variations of all material properties: 23 orders of magnitude between the conductivities of copper and sulphur as shown in Table 1.2.

1.3.3   Dielectric properties

How does a non-conducting material respond to the presence of an external electric field?

The dielectric constant or permittivity is a material property which relates the amount of electric polarization (charge displacement) of a material P under the action of an electric field ξ

$$\varepsilon = \varepsilon_0 \varepsilon_r = \frac{P}{\xi}.$$

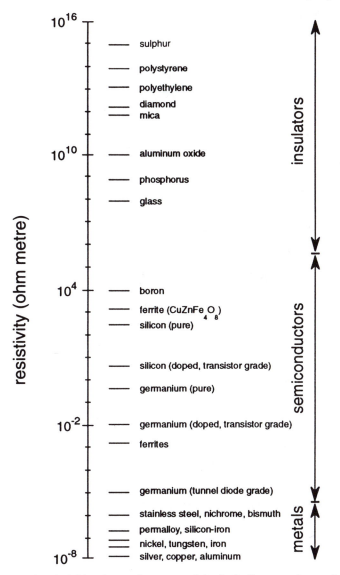

Fig. 1.2 Range of resistivities for various materials, including metals, semiconductors and insulators.

This means that a material with high permittivity gives a large electric polarization for a given field strength. Values of the relative permittivity can be as high as 7000 in barium titanate, but in most cases are much lower, for example the relative permittivity of water is 80 (Table 1.3). The relative permittivity of a material can also be determined from the capacitance C of a condenser

Table 1.2   Numerical values of electrical resistivities (ρ) and conductivities (σ) of various materials at room temperature

Material	ρ ($\Omega\,m$)	σ ($\Omega^{-1}\,m^{-1}$)
Sulphur	2.0×10^{15}	5.0×10^{-15}
Diamond	1.0×10^{12}	1.0×10^{-12}
Glass	1.0×10^{8}	1.0×10^{-8}
Boron	1.8×10^{4}	6.0×10^{-5}
Silicon	3.5×10^{2}	2.9×10^{-3}
Germanium	1.0×10^{-5}	0.1×10^{6}
Gadolinium	1.4×10^{-6}	0.7×10^{6}
Mercury	1.0×10^{-6}	1.0×10^{6}
Stainless steel	0.75×10^{-6}	1.3×10^{6}
Zinc	5.9×10^{-8}	16.9×10^{6}
Copper	1.7×10^{-8}	58.0×10^{6}
Silver	1.6×10^{-8}	63.0×10^{6}

Table 1.3   Dielectric properties of various materials

Material	ε_r
Copper	∞
Barium titanate	7000
Distilled water	80
Inorganic glasses	6–20
Alumina	9
Polyester	4
Polystyrene	2
Air	1
Vacuum	1

with the material as dielectric, compared with that of the same condenser C_0 with a vacuum in place of the material

$$\varepsilon_r = \frac{C}{C_0}.$$

The dielectric strength is a material property which represents the resistance of a material to electrical breakdown (i.e. spontaneous electrical conduction) under the action of a strong electric field. It is sometimes called the breakdown strength, and is measured in volts per metre. Below this field strength the material is an insulator, and above it, it is a conductor. Unfortunately this property varies widely even among materials which are nominally identical. Therefore it is not reliable to quote values of dielectric strength for particular materials, although they are typically in the range of $10^6\,V\,m^{-1}$.

1.4 OPTICAL PROPERTIES

How do we quantify the optical behaviour of materials?

The optical properties of a material tell us how the material interacts with incident electromagnetic waves. These properties can be expressed in terms of two optical constants. Often the refractive index n and the extinction coefficient k are used, both of which change with the wavelength of the incident light. Alternatively, we can define the optical properties using the reflectance R together with one of the above. We can also use instead the real and imaginary components of the dielectric constant ε. These quantities are the principal optical properties of interest, and all change with the frequency of the incident electromagnetic waves.

1.4.1 Refractive index

How does the speed of light in a material determine its change of direction at an interface?

The refractive index of a material is the ratio of wavelength, or phase velocity, of light in a vacuum to that in the material. It is a material property which can be obtained in principle solely from the measurement of the speed of light in a material, although this is never attempted in practice

$$n = \frac{\text{speed of light in vacuum}}{\text{speed of light in material}}.$$

The refractive index of a transparent material is usually determined on the basis of the measurement of two angles. θ_i is the angle of incidence of a light beam at the surface of the material and θ_r is the angle of refraction of the light beam inside the material

$$n = \frac{\sin \theta_i}{\sin \theta_r}.$$

In fact the refractive index is frequency dependent, which is why a prism can be used to split white light into different colours (dispersion on the basis of frequency).

1.4.2 Extinction coefficient k

How is light energy absorbed by a material?

The optical extinction coefficient k is defined as the fractional rate of decrease of light intensity dI/I in a material per unit path length multiplied by $\lambda/4\pi$

Table 1.4 Optical properties of various materials

Material	n	k	R
Aluminium	1.212	12.464	0.9697
Cobalt	4.46	5.86	0.722
Copper	0.44	8.48	0.976
Gold	0.13	8.03	0.992
Iron	3.43	4.79	0.678
Nickel	3.06	5.74	0.753

where λ is the wavelength

$$k = \frac{-\lambda}{4\pi I}\left(\frac{dI}{dx}\right).$$

This is a dimensionless material property. It is also dependent on the frequency of light.

We can also define an attenuation coefficient α which represents the rate at which the intensity of light decays with depth in a material

$$\alpha = -\frac{1}{I}\frac{dI}{dx} = \frac{4\pi k}{\lambda}.$$

1.4.3 Reflectance

How do we quantify the amount of light reflected at an interface?

The optical reflectance R is the fraction of incident light that is reflected from a surface. The value of R is dependent on both the frequency of the light and the angle of incidence

$$R = \frac{\text{reflected intensity}}{\text{incident intensity}}.$$

It is usually measured using normal incidence of light.

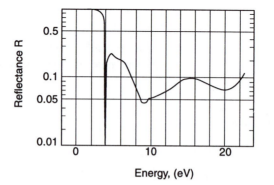

Fig. 1.3 Reflectance spectrum of polycrystalline silver [5]. © IEEE 1965

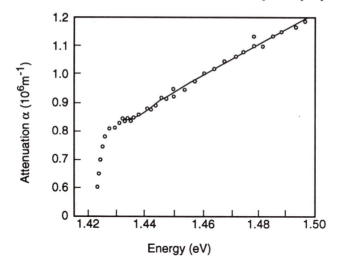

Fig. 1.4 Optical attenuation spectrum $\alpha(E)$ of gallium arsenide [6]. Reproduced with permission from M. D. Sturge, *Phys. Rev.*, 1962.

The optical constants shown in Table 1.4 are valid at an energy of 1 eV or equivalently at a wavelength of 1240×10^{-9} m.

The optical reflectances of metals and semiconductors have very characteristic features. Metals have high reflectance at long wavelengths but at shorter wavelengths the reflectance declines. On the other hand semiconductors have low reflectance at long wavelengths but beyond a threshold wavelength known as the band edge or absorption edge, the reflectance increases rapidly as the wavelength decreases. This frequency or energy dependence of optical properties is demonstrated in Figures 1.3 and 1.4.

1.4.4 The Hagen–Rubens law

Is there a relationship between the electrical and the optical properties of a metal?

The optical reflectivity and the electrical conductivity of metals at 'low' frequencies ($v < 1 \times 10^{14}\,\mathrm{s}^{-1}$) or long wavelengths ($\lambda > 3\,\mu\mathrm{m}$) are also related [7] by an equation of the form,

$$R = 1 - 2\sqrt{\frac{4\pi\varepsilon_0 v}{\sigma_0}}$$

where σ_0 is the DC electrical conductivity and R is the reflectance. This is known as the Hagen–Rubens relation [8]. Therefore the mechanisms underlying conductivity and reflectivity seem to be related. The prediction of the reflectance on the basis of the Hagen–Rubens law is shown in Fig. 1.5.

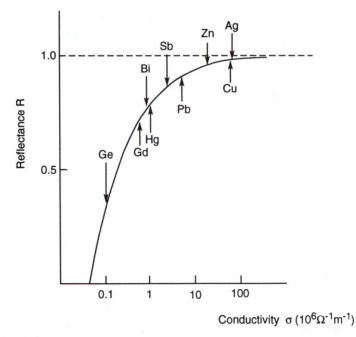

Fig. 1.5 Variation of the optical reflectance with electrical conductivity at a wavelength of 3 μm ($v = 1 \times 10^{14}\,\text{s}^{-1}$).

1.5   THERMAL PROPERTIES

How do we quantify the thermal properties of materials?

In the case of thermal properties we are often concerned with the rate of flow of heat per unit temperature gradient through the material as measured by the thermal conductivity K. This determines whether the material is a thermal conductor or insulator. Another quantity of interest is the amount of heat which must be supplied to raise the temperature of unit mass by one degree, that is the specific heat or heat capacity C.

1.5.1   Thermal conductivity

How is the thermal conductivity defined?

The thermal conductivity K of a material is the rate of transfer of heat per unit time, per unit cross-sectional area, per unit temperature gradient

$$K = \frac{Qx}{(T_2 - T_1)At} = \frac{J_Q}{(\mathrm{d}T/\mathrm{d}x)}.$$

Table 1.5 Thermal conductivities (K) of various materials

Material	$K\ (W\,m^{-1}\,K^{-1})$
Aluminium	0.0237
Copper	0.0398
Gold	0.0315
Iron	0.0080
Silicon	0.0083
Glass	0.00002

A is the cross-sectional area through which the heat passes, Q is the heat energy transferred in time t between two locations a distance x apart, where T_2 and T_1 are the temperatures at the two locations. An alternative, but equivalent definition is that K is the quotient of the thermal flux J_Q with respect to the temperature gradient dT/dx. This equation only applies under steady state conditions. Representative values of K are given in Table 1.5.

1.5.2 The Wiedemann–Franz law

Is there a relationship between the electrical and thermal properties of a metal?

In most cases good electrical conductors are also good thermal conductors. Quantitative investigation by Wiedemann and Franz [9] revealed that for most metals the relationship between electrical conductivity σ and thermal conductivity K obeyed the following law

$$\frac{K}{\sigma T} = 2.4 \times 10^{-8}\,J\,\Omega\,K^{-2}\,s^{-1}.$$

This seems to imply that the underlying mechanisms behind electrical and thermal conductivity are related in some way.

1.5.3 Specific heat capacity

What determines the increase in temperature of a material when it is heated?

The specific heat C_m of a material is the amount of heat required to raise a unit mass of the substance by one degree of temperature, while the heat capacity C is the amount of heat required to raise the temperature of an unspecified mass by one degree of temperature.

$$C_m = \frac{1}{m}\frac{dQ}{dT}$$

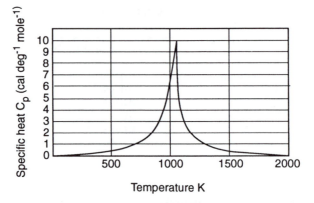

Fig. 1.6 Heat capacity of iron showing anomalous behaviour at about 1040 K [10]. Reproduced from *J. Phys. Chem. Sol.*, **1**, J. A. Hoffmann *et al.*, p. 52, copyright 1956, with kind permission from Pergamon Press Ltd, Oxford, UK.

$$C = \frac{dQ}{dT} = \frac{dU}{dT}.$$

Here m is the mass, dT is the change in temperature, U is the internal energy and dQ is the heat energy absorbed. The specific heat is itself dependent on temperature. It can also be dependent on whether the measurement is made under constant volume or constant pressure conditions.

The heat capacity of some materials varies in a very characteristic way. For example the temperature dependence of the heat capacity of iron which is shown in Fig. 1.6 has an anomaly at 1040 K. As we shall see later this also corresponds to a magnetic phase transition.

1.5.4   The Dulong–Petit law

Is there a relationship between the heat capacities of various materials?

The heat capacities of many materials are found to be linearly dependent on the molecular or atomic weight of the substance, at least at higher temperatures. This can be expressed as the 'molar heat capacity'. That is the heat capacity of a fixed number ($N_0 = 6.02 \times 10^{23}$) of atoms or molecules of a substance. For many materials this has a value close to 25 J mol^{-1} K^{-1}, a result discovered by Dulong and Petit [11] and shown in Table 1.6.

This seems to imply that the heat capacity is dependent only on the number of elementary entities, either atoms or molecules depending on the material. However even this law only applies at high temperatures, since the heat capacity varies with temperature as we shall see in the next chapter. The molar heat capacity deviates significantly from the value predicted by the Dulong–Petit law at lower temperatures (e.g. below 100 K in lead, below 400 K in aluminium and copper).

Table 1.6 Heat capacities (C) of various materials

Material	$C \, (J \, mol^{-1} \, K^{-1})$
Aluminium	24.3
Iron	25.7
Nickel	26.8
Copper	24.4
Lead	26.9
Gold	25.5

1.6 MAGNETIC PROPERTIES

How do we quantify the magnetic behaviour of materials?

When dealing with magnetic properties, we are usually concerned either with the permeability μ of the material, which describes its response to an external field, or its magnetization M which is the magnetic moment per unit volume.

1.6.1 Magnetic moment and magnetization

Which properties determine the response of a material to an applied magnetic field?

The magnetic moment m of a material is the maximum torque τ_{max} experienced by the material under the action of a magnetic field in free space, divided by the strength of the magnetic field H

$$m = \frac{\tau_{max}}{\mu_0 H}$$

where μ_0 is the permeability of free space which is a universal constant. The magnetization M of a magnetic material is the magnetic moment per unit volume

$$M = \frac{m}{V}.$$

This is often a function of magnetic field strength H, and so is not a material constant.

1.6.2 Magnetic susceptibility

How does a material respond to an external magnetic field?

The magnetic susceptibility χ is the ratio of magnetization M to magnetic field strength H.

Table 1.7 Magnetic permeability of various materials

Material	Relative permeability (μ_r)	Susceptibility (χ)
Permalloy	$\sim 10^4$	$\sim 10^4$
Iron	$\sim 10^2$–10^3	$\sim 10^2$–10^3
Samarium–cobalt	~ 1	–
Aluminium	1.00002	2.0×10^{-5}
Manganese	1.00083	8.3×10^{-4}
Copper	0.99999	-1×10^{-5}
Bismuth	0.99983	-1.7×10^{-4}

$$\chi = \frac{M}{H}.$$

It is not a material constant for strongly magnetic materials such as iron, cobalt and nickel because in these materials χ is strongly dependent on field strength H. However it can be approximately constant for weakly magnetic materials such as aluminium and sodium. Values of the susceptibility for various materials are shown in Table 1.7.

1.6.3 Magnetic permeability

How is the magnetic induction in a material related to the magnetic field?

The magnetic permeability of a material is the ratio of magnetic induction B to the magnetic field strength H

$$\mu = \frac{B}{H}$$

and since $B = \mu_0(H + M)$, there is an exact relationship between permeability and susceptibility

$$\mu = \mu_0(\chi + 1) = \mu_0\mu_r$$

where μ_r is the relative permeability and μ_0 is the permeability of free space.

This quantity is strongly dependent on magnetic field strength applied to ferromagnetic materials and so is not a material constant. In weakly magnetic materials such as paramagnets or diamagnets it is close to the value μ_0 since χ is approximately zero.

1.6.4 The Curie–Weiss law

How does the susceptibility of weakly magnetic materials vary with temperature?

Some magnetic materials undergo a magnetic phase transition at a temperature T_c from a high-temperature state (with low susceptibility $\chi \approx 10^{-3}$) to a low-temperature state (with a significantly higher susceptibility). In many cases the variation of the susceptibility of these materials with temperature T in the region $T > T_c$ can be described by an equation known as the Curie–Weiss law [12]

$$\chi(T) = \frac{C}{T - T_c}.$$

In other cases the value of T_c is zero, which means that there is no transition temperature, and in this case we have the more restricted Curie law [13].

$$\chi(T) = \frac{C}{T}.$$

In both cases the susceptibility varies inversely with temperature. An example of the variations of χ^{-1} with temperature for some manganese compounds, which obey the Curie or Curie–Weiss laws, are shown in Fig. 1.7.

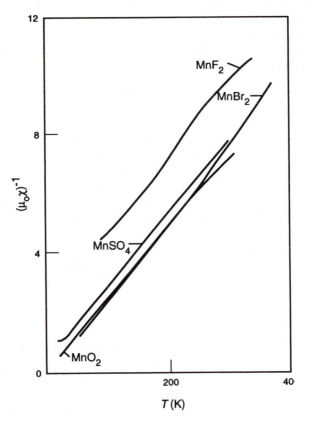

Fig. 1.7 Magnetic susceptibility of some manganese compounds as a function of temperature, showing agreement with the Curie law or the Curie–Weiss law.

1.7 RELATIONSHIPS BETWEEN VARIOUS BULK PROPERTIES

Are there any relationships between the bulk properties of materials which are not obvious on the basis of the continuum model?

In addition to the variation of the bulk properties of materials with temperature and applied fields, empirical relationships have been noticed between various macroscopic properties. We have considered five well known empirical laws determined from macroscopic properties which require an explanation based on a better understanding of the structure of materials. We have observed for example that a material with high optical reflectance is usually a good electrical conductor and is also usually a good thermal conductor. This and other property relationships are largely unexplained by the continuum model. These relationships cannot be simply fortuitous and they therefore provide the starting point for the development of a theory of the material properties because they challenge us to provide an explanation.

We need to understand the underlying physical reasons why such relationships occur. It is common knowledge that good optical reflectors are good electrical conductors. Despite the everyday familiarity of this relationship, it is at first a strange result because it is difficult to see why there should be any relation between these two properties at all.

1.8 CONCLUSIONS

The main objective of this chapter has been to indicate the types of macroscopic measurement that are usually made for property determination of materials. These measurements are almost always macroscopic measurements which do not depend on any assumptions about the underlying mechanisms, nor do they depend on a more fundamental interpretation. They are merely defined in terms of an individual measurement or sometimes in terms of a combination of measurements.

Our inability to explain these observations through the use of the continuum model forces us to develop more sophisticated descriptions of materials based on an atomistic approach, and ultimately on an electronic approach.

REFERENCES

1. Landau, L. D., Akhiezer, A. I. and Lifshitz, E. M. (1967) *General Physics: Mechanics and Molecular Physics*, Pergamon Press, Oxford.
2. Weast, R. C. (1985) *Handbook of Chemistry and Physics*, 66th En, CRC Press, Boca Raton, Florida.
3. Jiles, D. C. and Palmer, S, B. (1980) *J. Phys. F.*, **10**, 2857.
4. Ohm, G. S. (1827) *Die Galvanische Kette, Mathematisch Bearbeitet* (The Galvanic Circuit Investigated Mathematically), T.H. Riemann, Berlin (reprinted by Kraus Publishers, New York 1969).

5. Ehrenreich, H. *et al.* (1965) *IEEE Spectrum*, **2**, 162.
6. Sturge, M. D. (1962) *Phys. Rev.*, **127**, 768.
7. Longhurst, R. S. (1967) *Geometrical and Physical Optics*, 2nd Edn, Longmans, p. 477.
8. Hagen, E. and Rubens, H. (1903) *Ann. Phys.*, **11**, 873.
9. Wiedemann, G. and Franz, R. (1853) *Ann. Phys.*, **89**, 497.
10. Hofmann, J. A., Paskin, A., Tauer, K.J. and Weiss, R.J. (1956) *J. Phys. Chem. Sol.*, **1**, 45.
11. Dulong, P. L. and Petit, A.T. (1818) *Annales de Chemie et de Physique*, **7**, 122.
12. Weiss, P. (1908) *Compt. Rend.*, **143**, 1138.
13. Curie, P. (1895) *Ann. Chem. Phys.*, **5**, 289.

FURTHER READING

The following is an excellent series of articles on the physical properties of materials.

Cottrell, A. H. (1967) The nature of metals. *Scientific American*, **217**, 90.
Ehrenreich, H. (1967) The electrical properties of materials. *Scientific American*, **217**, 194.
Javan, A. (1967) The optical properties of materials. *Scientific American*, **217**, 238.
Keffer, F. (1967) The magnetic properties of materials. *Scientific American*, **217**, 222.
Ziman, J. M. (1967) The thermal properties of materials. *Scientific American*, **217**, 180.

EXERCISES

For each Exercise explain the well known law. You should describe the relations involved and try to explain the underlying resason for each on a classical basis.

Exercise 1.1   The Wiedemann–Franz law

Exercise 1.2   The Hagen–Rubens relation

Exercise 1.3   The Dulong–Petit law

2
Properties of Atoms in Materials

This chapter provides a background for the rest of the book since clearly the electrons within materials are contained within a volume defined by the atomic cores of the material, and it is important to realize that the properties of those electrons are determined largely by their interactions with these atomic cores. In fact we should go even further and state that the electronic properties of interest in this book are determined exclusively by the energy 'landscape' provided by the ionic cores. Therefore before proceeding we should look at some of the properties of this ionic background. In the case of crystalline materials this ionic background forms an ionic 'lattice' but in other materials, such as amorphous solids, it forms a random aggregate.

2.1 THE ROLE OF ATOMS WITHIN A MATERIAL

What is the next level of sophistication beyond the continuum model?

Moving on from the continuum model our next question must be to ask what lies beneath the surface of the material. Before proceeding with a discussion of the electronic properties of materials, we shall pause briefly to consider the arrangement of the atomic cores within a solid. The simplest case occurs when the atoms are arranged in a regular crystal lattice with a well defined spatial periodicity or symmetry.

Many authors like to consider the crystal structure to be the combination of a crystal lattice plus a 'basis' (which is an identical configuration of atoms attached to each lattice point) [1, p. 4]. We shall go further and consider our solids to be composed of a periodic array of atomic cores immersed in a sea of electrons.

$$\text{solid} = \text{ionic lattice} + \text{electronic 'sea'}.$$

The lattice here represents the arrangement of atoms in a periodic structure and the electron sea is a random arrangement of high-energy electrons without obvious periodicity, although we shall notice later that the lattice imposes periodicity on the electrons. This separation may seem almost trivial at first sight, but it does have important implications for our subsequent description of materials. Some of the properties of materials can be attributed principally to the lattice and some principally to the electrons. It will be our purpose to separate and identify these, where possible, and of course we will later pay more attention to those properties which depend on the electrons.

Since all electrons are identical we will eventually see that the differences in the electronic properties of materials are due more to the ions and their lattice symmetry than to the electrons themselves, which is an interesting conclusion.

2.1.1 Types of lattice symmetry

How do the atoms arrange themselves inside a material?

The classification of the various types of lattice is merely a mathematical abstraction of the forms of symmetry exhibited by materials [2,3]. Figure 2.1 shows the various forms of lattice or crystal structures. These have been verified by X-ray diffraction results [4].

The principal classes of lattice, with the number of elements exhibiting each form of symmetry in the solid state, are as follows: 35 cubic (19 face centred cubic, 14 body centred cubic, 2 simple cubic); 29 hexagonal (mostly hexagonal close-packed); 5 trigonal; 2 tetragonal; 21 orthorhombic, monoclinic or triclinic.

2.1.2 Cohesive energy of the lattice

What forces hold the lattice together?

When considering the structure of a solid two questions arise: (i) what holds the lattice together? and (ii) what determines the symmetry of the lattice? The cohesive energy of a crystal is the amount of energy which must be supplied in order to separate it into free neutral atoms. The melting temperatures of crystals are approximately proportional to the cohesive energy. Typical values of the cohesive energy of solids are shown in Table 2.1.

The cohesive energy of a solid is determined exclusively by electrostatic interactions between the electrons, which have negative charge, and the ionic cores, which have positive charge. The ionic cores are located on the lattice sites, whereas the electrons may be located on the lattice sites or may be free to move throughout the solid. The resulting variation in interatomic potential is shown in Fig. 2.2.

The exact form of the electrostatic interactions holding the solid together may be different in different cases. We can identify three main types:

1. ionic interactions, e.g. NaCl;
2. covalent bonds, e.g. carbon; and
3. van der Waals interactions in inert crystals, e.g. xenon.

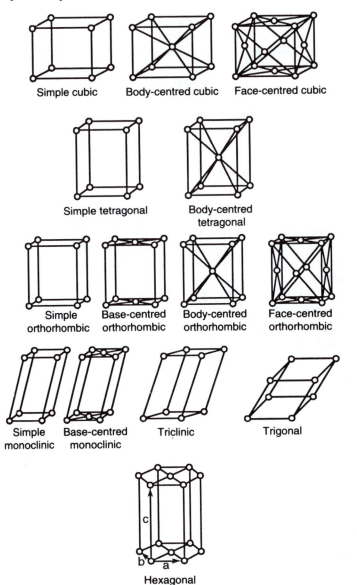

Fig. 2.1 Different forms of lattice symmetry exhibited by crystalline materials.

There is also the question of the interaction in metals in which all the ionic cores have a net positive charge and are immersed in a sea of negatively charged electrons. In these cases the negatively charged sea of electrons is composed of some of the outer electrons of the atoms, while the inner electrons remain localized at the ionic sites. The subject of the cohesive forces in materials has been treated in detail by Maitland *et al.* [5].

Table 2.1 Values of cohesive energies for various materials

Material	Cohesive energy E_c (kJ mol^{-1})	Melting point	
		T_c (K)	T_c (°C)
Ruthenium	650	2723	2450
Silicon	446	1687	1410
Iron	413	1811	1575
Germanium	372	1211	985
Cadmium	112	594	321
Mercury	65	234	− 39
Xenon	16	161	− 112
NaCl	777	1074	801

In cases where there is no separation of electrical charges within a material the electrostatic interactions cannot be very large. In these cases the atoms are held together by the 'fluctuating dipole forces' also known as van der Waals forces. These forces are dominant in the inert solids such as Ne, Ar, Kr, Xe, and Rn. They form the weakest type of bond and only play the most important role if none of the other cohesive mechanisms occurs.

In these solids, although the time average of the charge distribution is always zero for a given atom, at any given instant there will be a net electric dipole moment. The dipole moments on neighbouring lattice sites lead to a weak attractive interaction between the dipoles. This force holds the atoms together.

2.2 THE HARMONIC POTENTIAL MODEL

How can the atomic forces be modelled in the simplest way?

We now need to consider the form of bonding between the atoms in order to provide an adequate model for the behaviour of the lattice. The interactions between the individual ions can be represented by a single force. Over a large range of deformations the force between the neighbouring lattice sites is proportional to their displacement. Ultimately for large deformations this proportionality no longer holds of course, but we should first explore the consequences of this simplest of models.

In the case illustrated in Fig 2.3, the harmonic potential leads to an internal restoring force F on each atom which is proportional to the displacement u of the individual atom from its unstrained equilibrium position with respect to its two nearest neighbours [6],

$$F = -2ku$$

where k is a force constant. Since the force is $F = -dE_P/du$ where E_P is the

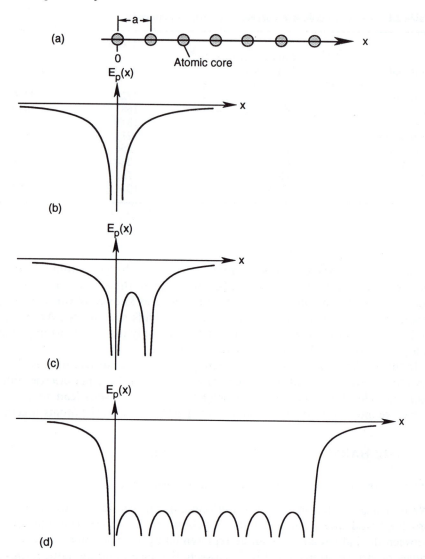

Fig. 2.2 Variation of the interatomic potential $E_p(x)$ of the ions with position in a regular crystal lattice: (a) linear lattice of ions; (b) potential well of isolated ions; (c) potential wells of two ions in proximity; and (d) periodic potential of a linear array.

potential energy, this means that each lattice site reaches its equilibrium position at the bottom of a parabolic, or harmonic, potential of the form

$$E = -\int (-2ku)\,du = ku^2.$$

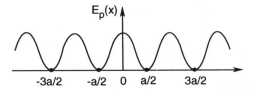

Fig. 2.3 Ionic cores in a one-dimensional crystal lattice subject to a harmonic potential.

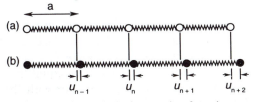

Fig. 2.4 Linear lattice of atoms in which the interaction force is represented by the spring model.

Such a potential can arise also from a mechanical model in which the atomic cores are connected by linear springs (Fig 2.4). In this case the force on any given atom in the lattice is determined solely by its displacement u relative to its two nearest neighbours

$$F = -ku \qquad + k(-u) \qquad = -2ku$$

force due to   force due to
atom on left   atom on right.

Consider now for simplicity a transverse vibration of the linear lattice with interactions only between nearest neighbours. For the nth atom in the lattice the force is

$$F = m\frac{d^2u_n}{dt^2} = k(u_{n-1} - 2u_n + u_{n+1})$$

where u_n is the displacement of the nth atom and m is its mass. Using a Taylor series expansion for the displacement

$$u_{n+1} = u_n + \left(\frac{du_n}{dx}\right)a + \left(\frac{d^2u_n}{dx^2}\right)\frac{a^2}{2} + \cdots$$

where a is the lattice spacing. Substituting this back into the force equation gives,

$$F = m\frac{d^2u_n}{dt^2} = ka^2\frac{d^2u_n}{dx^2}$$

which is the equation of a wave with velocity v given by

$$v = \sqrt{\frac{ka^2}{m}}.$$

This gives a direct relationship between the form of the interatomic potential through the force constant k, the lattice spacing a, the mass of the atoms and the velocity of vibrations passing through the lattice. We will see in the next section that the elastic modulus is also determined by these quantities.

2.2.1   Elastic modulus

Can a simple expression for the elastic modulus be derived on the basis of the spring model?

We can use the above simple linear lattice model, in which each atomic core lies within its harmonic potential, to calculate the elastic modulus E_Y. Suppose the lattice is subjected to a constant applied force F_{app} along its length. Under equilibrium conditions the lattice spacing will be equal throughout the chain, ensuring no net force on each atom in accordance with the above equations. Nevertheless this lattice spacing will be different from the unstrained spacing a. This equilibrium lattice spacing under the applied force will be $a + u_n - u_{n-1}$.

Considering the forces on the last atom in the lattice, since this is the only one acted upon directly by the applied force, it has only one nearest neighbour. This leads to a restoring force of $-k(u_n - u_{n-1})$ which under equilibrium conditions must balance the externally applied force. Also under equilibrium conditions we can assume that $u_n - u_{n-1}$ is the same for all pairs of atoms. Let this be Δu. For the purposes of this derivation we will assume that the cross-sectional area per linear lattice chain is a^2, which is equivalent to assuming a simple cubic lattice. This gives a stress of $\sigma = F/a^2$

$$\sigma = \frac{k\Delta u}{a^2}.$$

Let the macroscopic deformation of the lattice be Δl. The strain e is then

$$e = \frac{\Delta l}{l_0} = \frac{N\Delta u}{Na} = \frac{\Delta u}{a}$$

where N is the total number of atoms in the linear lattice and l_0 is the undeformed macroscopic length. The elastic modulus E_Y is the ratio of stress to strain, and therefore

$$E_Y = \frac{\sigma}{e} = \frac{k}{a}.$$

This means that the elastic modulus of the linear chain lattice is defined by the unstrained interatomic spacing a and the force constant k. The elastic modulus of the lattice can then be related to the wave velocity in the lattice by the equation

$$v = \sqrt{\frac{E_Y a^3}{m}} = \sqrt{\frac{E_Y}{\rho}}$$

where ρ is the density of the material.

We should note that under large deformations this simple parabolic potential is not sufficient to correctly describe the elastic behaviour, and in this case higher order terms become significant, particularly third-order contributions to the elastic modulus. This leads to anharmonicity of the lattice potential which is essential in the explanation of such phenomena as thermal expansion.

2.2.2 Quantization of lattice vibrations

If the material consists of a discrete lattice are all frequencies of vibration allowed?

We consider again the simplified problem of the linear lattice. The expression for the force on an individual atom can be transformed into an equation in terms of the positions of the atoms x instead of their displacements from equilibrium u.

Consider a longitudinal vibration of the lattice. Let x_n be the position of the nth atom. Then the displacement of the nth atom from unstrained equilibrium can be written $u_n = x_n - (x_{n-1} + a)$ or alternatively $u_n = x_n - (x_{n+1} - a)$. The force on the nth atom is then

$$F_n = -k(x_n - (x_{n-1} + a)) + k(-x_n + (x_{n+1} - a)).$$

At any lattice spacing the nth atom will be at equilibrium, according to this model, provided it is located midway between its nearest neighbours. Therefore the net force on the atom is not dependent on the equilibrium spacing a. Eliminating a leads to

$$F = k(x_{n-1} - x_n) - k(x_n - x_{n+1}).$$

Then the force and acceleration are related by the equation

$$F = m\frac{d^2x_n}{dt^2} = k(x_{n-1} - 2x_n + x_{n+1}).$$

This again leads to the following equation for the velocity of propagation of vibrations

$$v = \sqrt{\frac{ka^2}{m}} = \sqrt{\frac{E_Y}{\rho}}$$

For the position of the nth atom in the chain this equation has solutions of the form

$$x_n = A\exp(i(qna - \omega t)).$$

This is a valid solution of the equation of motion for any value of A provided the following condition is satisfied for the frequency of vibration ω

$$\omega^2 = \frac{2k}{m}(1 - \cos qa).$$

From this we may take the positive root

$$\omega = \sqrt{\frac{4k}{m}} \sin\left(\frac{qa}{2}\right).$$

If we have a total of N atoms in the chain and impose periodic boundary conditions (the Born–von Karman condition), then there are just N allowed values of q which independently satisfy all the conditions [6].

The proof of this is fairly simple. If we have a wave passing along the lattice we must have $x_N = x_1$ for periodic boundary conditions. Therefore the allowed values of the product qNa must be an integral multiple of 2π in order to satisfy the periodic boundary condition

$$qNa = 2\pi, 4\pi, 6\pi, \ldots\ldots 2N\pi.$$

Since a is the interatomic spacing and N is the total number of atoms, then $Na = L$, the length of the lattice. Therefore

$$q = 2\pi/L, 4\pi/L, \ldots\ldots 2N\pi/L.$$

Let us try a larger value of q which can still meet the boundary conditions, for example $q = 2(N + 1)\pi/L$, in order to see what happens when we go outside this specified range of q.

$$x_n = A \exp i \left\{ \frac{2(N + 1)\pi}{L} na - \omega t \right\}$$

$$= A \exp i \left\{ \frac{2\pi}{L} na - \omega t \right\} \exp i \left\{ \frac{2N\pi}{L} na \right\}$$

and since $N/L = a$, this gives the followng relation for all n,

$$x_n(2(N + 1)\pi) = x_n(2\pi) \exp i \{2n\pi\}$$

which is not an independent solution. Higher values of q will simply lead to wave motion that is identical to one of the values of q in the range $(2\pi/L, 4\pi/L, \ldots, 2N\pi/L)$. In our linear lattice on N atoms there are consequently only N allowed distinct vibrational modes [7]. This is demonstrated graphically in Fig. 2.5.

This can easily be extended to three dimensions. Consider an array of N^3 atoms in a cubic lattice, the allowed values of q are

$$q = q_x + q_y + q_z$$

where now the constraint is

$$q_x; q_y; q_z = 2\pi/L, 4\pi/L, \ldots\ldots 2N\pi/L.$$

Fig. 2.5 Waves in a discrete lattice showing that two different values of wave vector give identical displacements of the lattice. The longer wavelength vibration is the only meaningful interpretation of the wave on the basis of the discrete lattice. This demonstrates that there are only N independent solutions of the wave equation in a lattice with N atoms.

Again only certain vibrational modes are possible and there are N^3 distinct modes. We have therefore shown by this simple example that the allowed vibrations of a periodic lattice are restricted and discrete, that is they are quantized. These quantized vibrations are called phonons. This is an important result in which we can understand the quantization of the allowed lattice vibrations on the basis of a simple discrete classical model of the material.

2.2.3 Anharmonicity

What are the immediate and obvious drawbacks of the simple spring model?

Although the harmonic potential, or spring model, works quite well for small displacements of the atoms, it is quite easy to demonstrate that it must fail for large displacements. If we simply consider two atoms, the energy of the system when the atoms are moved closer together will be larger than the energy of the system if they are moved apart. Put another way, the atoms cannot be displaced relative to one another so that they occupy the same location, but they can have the distance between them doubled.

This argument demonstrates that the lattice potential must ultimately be anharmonic [8]. It is this anharmonicity which gives rise to thermal expansion and is also responsible for third-order elastic constants, that is the variation of elastic moduli with strain. The anharmonic potential can most easily be described as a perturbation from our simple harmonic potential

$$E_P = ku^2$$

by the addition of third- and fourth-order terms

$$E_P = ku^2 - fu^3 - gu^4$$

where all the coefficients k, f and g are positive.

Alternatively you will often find the anharmonic potential expressed as a Lennard-Jones '6–12' potential [9] which has the form shown in Fig. 2.6

$$E_P = -\frac{A}{a^6} + \frac{B}{a^{12}}$$

Energy E

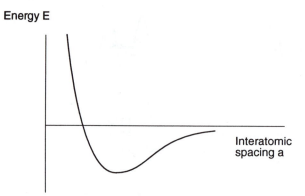

Interatomic
spacing a

Fig. 2.6 Anharmonic potential as a function of distance between two ionic cores.

where a is the lattice spacing. However this is also just a simple method of obtaining an approximate form of energy as a function of interatomic spacing.

2.3   SPECIFIC HEAT CAPACITY

Can the observed specific heat capacity be interpreted in terms of the lattice properties?

As we consider the physical properties of materials and try to explain their behaviour we will find that these properties often can be attributed primarily to either the lattice or the electron sea. The first of these which we will consider is the heat capacity or the specific heat, which is the heat capacity per unit mass. This is determined largely by the lattice, although there is a contribution due to the electrons which we shall discuss later.

The vibrations of a crystal lattice are related to the heat content of the system and hence to the thermodynamic temperature. We have already noted the Dulong–Petit law which simply states that the molar heat capacity is a constant. Let us see how this can be obtained from a consideration of lattice vibrations.

2.3.1   Classical theory of heat capacity

How can the heat capacity of a material be explained in terms of the vibration of the atoms?

From classical statistical thermodynamics we expect a thermal energy $k_B T$ to be associated with each mode of vibration at any given temperature T. If there are N atoms each with three degrees of freedom, then there are $3N$ modes each with energy $k_B T$. The internal energy is therefore

$$U = 3Nk_B T$$

and since the heat capacity at constant volume is simply $C_v = dU/dT$, then

$$C_v = 3Nk_B.$$

If we consider a mole of material, for which $N = 6.02 \times 10^{23}$ atoms, and $k_B = 1.38 \times 10^{-23}\,\text{J K}^{-1}$ we obtain

$$C_v = 24.94\,\text{J mol}^{-1}\,\text{K}^{-1}$$

which is the classical value of the molar heat capacity predicted by the Dulong–Petit law.

Notice however from Fig. 2.7 that the heat capacity is not constant as temperature changes. Therefore a more sophisticated theory is needed, particularly at lower temperatures.

In the quantum theory of heat, atoms can only take up energy in discrete amounts (quanta) rather than continuously, as we have assumed so far. At room temperature the magnitude of the energy quanta that most atoms can absorb is small enough that these can be provided by the thermal vibrations of the surroundings. Exchange of heat with the surroundings can occur almost continuously leading to equipartition of energy and this means that these atoms have the classically expected heat capacity. Examples of materials that come

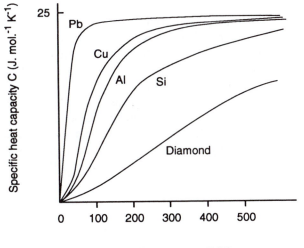

Fig. 2.7 Variation of specific heat C_v of various materials as a function of temperature. Reproduced with permission from R. E. Hummel, *Electronic Properties of Materials*, 2nd edn, published by Springer Verlag, 1993.

into this category at room temperature are lead, copper, aluminium and the other materials given in Table 1.6.

However in the case of materials which have light, strongly bound atoms, such as diamond, the energy quanta that must be absorbed are much larger. This means that exchange of heat with the surroundings is no longer continuous and so equipartition of energy is not established at room temperature. The heat capacity is therefore lower than the classically expected value.

For sufficiently low temperatures equipartition of energy breaks down for all materials. This occurs when the available thermal energy quanta become comparable to the lattice vibration quanta or phonon energies. Therefore, at sufficiently low temperatures the heat capacities of lead, copper, aluminium and other metals are reduced as shown in Fig. 2.7 and they no longer obey the classical Dulong–Petit law. In fact, at low temperatures the heat capacity obeys a relation of the form,

$$C_v = 233.8\, N_0 k_B \left(\frac{T}{\theta_D}\right)^3 \text{J mol}^{-1}\,\text{K}^{-1}$$

which is known as Debye's law. In this law θ_D is a characteristic temperature known as the Debye temperature, which varies from material to material.

2.3.2 Quantum corrections to the theory of heat capacity

How can we account for the deviations from the Dulong–Petit law at low temperatures?

We will now consider a correction to the theory of heat capacity which incorporates quantum effects. This we will term the Einstein–Debye theory in recognition of the two main contributors. The original ideas of the single-frequency Einstein theory were extended and generalized by Debye to give an equation which applies across the entire temperature range.

Einstein showed that the lattice vibrations of crystalline solids should be quantized and that therefore only certain vibrational modes were allowed. These vibrational quanta are called phonons. If we consider a single oscillator, that is an atomic core, oscillating at frequency ω, the average number of phonons N at a given temperature T obeys following equation [10]

$$N = \frac{1}{\exp\left(\dfrac{\hbar\omega}{k_B T}\right) - 1}$$

where $\hbar$ is Planck's constant divided by 2π (which has the value 1.054×10^{-34} J s), $\hbar\omega$ is the phonon energy, and $k_B T$ is the thermal energy available to each vibrational mode at a temperature T, including both kinetic and potential energy. For a single frequency of oscillation therefore the thermal energy of the material is

$$U = N\hbar\omega$$

$$U = \left(\frac{1}{\exp(\hbar\omega/k_B T) - 1}\right)\hbar\omega.$$

Considering the thermal energy of a solid, and remembering that a mole of material contains N_0 atoms and hence $3N_0$ oscillators, the thermal energy per mole is

$$U = 3N_0 \frac{\hbar\omega}{\exp\left(\dfrac{\hbar\omega}{k_B T}\right) - 1}.$$

Consequently according to the Einstein theory [11], the molar heat capacity should be

$$C_v = 3N_0 k_B \left(\frac{\hbar\omega}{k_B T}\right)^2 \frac{\exp\left(\dfrac{\hbar\omega}{k_B T}\right)}{\left(\exp\left(\dfrac{\hbar\omega}{k_B T}\right) - 1\right)^2}.$$

A graph of the prediction of the Einstein theory of heat capacity is given in Fig. 2.8.

Note that for high temperatures we obtain $\exp(\hbar\omega/k_B T) \simeq 1 + \hbar\omega/k_B T$, giving $C_v \simeq 3N_0 k_B$, which is the classical limit, expressed by the Dulong–Petit law.

The Einstein model of heat capacity works well at high and intermediate temperatures. It predicts a heat capacity which decreases as the temperature is reduced, which is in accordance with experimental observation. But the decrease that is observed in practice is not quite as rapid as the Einstein model suggests.

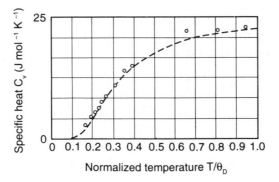

Fig. 2.8 Comparison of the prediction of the Einstein model with observed values of heat capacity. Reproduced with permission from C. Kittel, *Introduction to Solid State Physics* 6th edn, published by John Wiley & Sons, 1986.

The theory therefore does not work well at low temperatures and so a further correction to the theory is needed at these temperatures.

2.3.3   Extended theory of lattice vibrations

Can the single-frequency Einstein theory be improved to give better agreement with observations at low temperature?

In the Einstein model only one frequency of vibration ω was considered. If we allow interactions between the atoms, many more frequencies can exist which range from the Einstein frequency down to frequencies in the acoustic range. This extension to the theory was developed by Debye [12].

It is then necessary to know the number of atoms vibrating with any given frequency ω. This is the vibrational density of states $D(\omega)$ and in three dimensions this is given by [1, p. 106],

$$D(\omega) = \frac{3V\omega^2}{2\pi^2 v^3}$$

where V is the volume of the specimen and v is the velocity of sound which is assumed constant. The total energy of vibration of the solid is then simply

$$U = \int U_{\text{osc}}(\omega)\, D(\omega)\, d\omega$$

where $U_{\text{osc}}(\omega)$ is the energy of one oscillator at frequency ω which is given by the same expression as before

$$U_{\text{osc}}(\omega) = \frac{\hbar\omega}{\exp\left(\dfrac{\hbar\omega}{k_{\text{B}}T}\right) - 1}.$$

We now need to integrate this energy over the range of allowed vibrational modes. Substituting the expressions for U_{osc} and $D(\omega)$ into the above equation for the total energy of vibration gives

$$U = \frac{3V\hbar}{2\pi^2 v^3} \int_0^{\omega_{\text{D}}} \frac{\omega^3}{\exp\left(\dfrac{\hbar\omega}{k_{\text{B}}T}\right) - 1}\, d\omega$$

where ω_{D} is the so called 'Debye frequency' above which the oscillators behave classically. If we wish, we can define a 'Debye temperature' θ_{D} in terms of this Debye frequency

$$\theta_{\text{D}} = \frac{\hbar\omega_{\text{D}}}{k_{\text{B}}}.$$

2.3.4 Significance of the Debye temperature

How can we interpret the Debye temperature in terms of the properties of the lattice and the interactions?

The Debye temperature marks the boundary between the high-temperature classical behaviour which approximately follows Maxwell–Boltzmann statistics (and hence leads to the Dulong–Petit law) and a low-temperature region in which quantum statistics must be used.

Since the Debye temperature marks the boundary between the quantum and classical regimes, its value tells us something about the atomic bonding in the material we are dealing with. A high value of θ_D implies a lattice with light atoms and strong interactions between the atoms. For example, diamond has $\theta_D \simeq 2000$ K. On the other hand, lead which has heavy atoms which are weakly bound together has $\theta_D \simeq 100$ K.

The Debye temperature is in many ways analogous to the Fermi temperature in the theory of electron states which we will discuss in Chapter 4. It plays the same role in the theory of lattice vibrations as the Fermi temperature plays in the theory of electrons in metals.

2.3.5 Quantum theory of heat capacity

What is the heat capacity expected on the basis of the quantum theory of lattice vibrations?

For N atoms or ions the total phonon energy is

$$U = 9Nk_BT \left(\frac{T}{\theta_D}\right)^3 \int_0^{(\hbar\omega_D/k_BT)} \left(\frac{x^3}{e^x - 1}\right) dx$$

where

$$x = \frac{\hbar\omega}{k_BT}.$$

The Debye heat capacity is the derivative of the total energy of vibration U with respect to the temperature T

$$C_V = \frac{dU}{dT}.$$

The molar heat capacity is then

$$C_V = \frac{3V_0 h^2}{2\pi^2 v^3 k_B T^2} \int_0^{\omega_D} \frac{\omega^4 \exp\left(\dfrac{\hbar\omega}{k_B T}\right)}{\left[\exp\left(\dfrac{\hbar\omega}{k_B T}\right) - 1\right]^2} \, d\omega$$

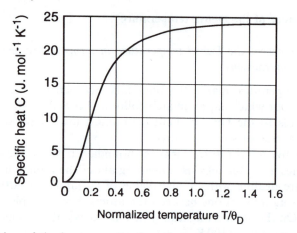

Fig. 2.9 Variation of the heat capacity C_v with temperature according to the Debye theory. Reproduced with permission from C. Kittel, *Introduction to Solid State Physics* 6th edn, published by John Wiley & Sons, 1986.

$$C_v = 9k_BN_0\left(\frac{T}{\theta_D}\right)^3 \int_0^{\theta_D/T} \frac{x^4 e^x}{(e^x - 1)^2}\, dx$$

where V_0 is the molar volume and N_0 is the number of atoms or ions in a mole. A plot of the Debye heat capacity C_v against temperature is given in Fig. 2.9.

2.3.6 Heat capacity at low temperatures: the Debye law

Can we obtain a simple expression for the heat capacity at low temperatures?

The heat capacity can be found from the above expression by evaluating the integral on the right-hand side. At low temperatures this results in the following expression for the heat capacity,

$$C_v = \frac{12\pi^4}{5}N_0k_B\left(\frac{T}{\theta_D}\right)^3$$

which gives the experimentally observed dependence of the heat capacity on T^3 at low temperatures known as the Debye T^3 law which was mentioned in section 2.3.1.

2.3.7 Heat capacity at high temperatures: the classical limit

What is the predicted heat capacity at high temperatures?

At high temperatures the integral has the value $x^3/3$ where $x = \theta_D/T$. So that,

$$C_v = 3N_0k_B$$

which of course is the expected classical Dulong–Petit result.

2.4 CONCLUSIONS

In this chapter we have discussed some of the properties of materials which arise principally from the behaviour of the atoms. We have also introduced the concept of quantization of allowed vibrational energies in discrete lattices. This serves as a precursor to the discussion of quantization of electron energy levels in Chapters 4, 5 and 6. It has given us a relatively simple introduction to the concept of quantized energy states because it can be quite easily visualized, even from a classical argument, why the quantization of lattice vibration occurs.

When we come to discussing the quantization of electron energies rather more abstraction will be required. However, having understood the reasons for quantization of lattice vibrations in this chapter, it should be easier to follow the discussion of quantization of electron energies in which the concept of imposed boundary conditions again plays a crucial role.

REFERENCES

1. Kittel, C. (1986) *'Introduction to Solid State Physics'*, 6th edn, Wiley, New York.
2. Pearson, W. B. (1972) *'Crystal Chemistry and Physics of Metals and Alloys'*, Wiley Interscience, New York.
3. Megaw, H. D. (1973) *'Crystal Structures: A Working Approach'*, W. B. Saunders, Philadelphia.
4. Cullity, B. D. (1978) *'X-Ray Diffraction'*, 2nd edn, Addison-Wesley, Reading, Mass.
5. Maitland, G. C., Rigby, M., Smith, E. B. and Wakeham, W. A (1981) *'Intermolecular Forces: Their Origin and Determination'*, Clarendon Press, Oxford.
6. Born, M. and Huang, K. (1954) *'Dynamical Theory of Crystal Lattices'*, Clarendon Press, Oxford.
7. Cochran, W. (1973) *'Dynamics of Atoms in Crystals'*, Crane & Russak.
8. Cowley, R. A. (1968) *Rep. Prog. Phys.*, **31**, 123.
9. Lennard-Jones, J. E. and Ingham, A. E. (1925) *Proc. Roy. Soc.*, **A107**, 636.
10. Zemansky, M. W. (1968) *'Heat and Thermodynamics'*, 5th edn, McGraw-Hill, New York, p. 312.
11. Einstein, A. (1907) *Ann. der Physik*, **22**, 180.
12. Debye, P. (1912) *Ann. der Physik*, **39**, 789.

FURTHER READING

Anderson, J. C. and Leaver, K. D. (1969) *'Materials Science'*, Van Nostrand Reinhold, New York.
Rosenberg, H. M. (1978) *'The Solid State'*, 2nd edn, Oxford University Press, Chapters 5 and 6.

EXERCISES

Exercise 2.1   Elastic modulus of a linear atomic lattice. The potential energy of a pair of atoms in a crystal is of the form

$$E_P(a) = \alpha_1 a^{-9} - \alpha_2 a^{-1}$$

where *a* is the interatomic separation. The equilibrium separation is 0.3 nm and the cohesive energy is 4eV (equivalent to 386 kJ mol^{-1}). Find the effective modulus of elasticity for the pair of atoms and determine the force which would be needed to reduce the spacing by 1%.

Exercise 2.2   Lattice stabilized by electrostatic repulsion. Consider a one-dimensional lattice made up of atoms with a charge of 1.6×10^{-19} Coulomb and a mass of 107×10^{-27} kg each. The array is held together by electrostatic forces with neighbouring atoms at a distance of 0.5 nm. Estimate the velocity of a long wavelength lattice vibration and calculate the elastic Young's modulus. (You may assume that the force between the ions is given by Coulomb's law $F = q_1 q_2 / 4\pi\varepsilon_0 x^2$ where x is the separation between the charges q_1 and q_2.)

Exercise 2.3   Classical and Debye theories of specific heat. Use the Dulong–Petit relation to calculate the classically expected thermal energy of 1 gram mole of material at 300 K. Aluminium has a Debye temperature of 430 K. Prove that the Debye theory and the Dulong–Petit law give the same results at high temperatures. Estimate the thermal energy of 1 gram mole of aluminium at 300 K by using Fig. 2.9. Explain why the results are different.

3

Conduction Electrons in Materials – Classical Approach

In this chapter we shall approach the description of electrons in solids using one of the simplest models possible, that of electrons as classical particles moving almost freely within the material experiencing minimal interactions with the ionic potential. In fact the model assumes that the ionic potential is completely flat, and that the only constraints on electron motion, apart from electron–electron collisions, are provided by the physical boundary of the material. At first such a model seems so far from reality as to be probably of little use, however quite surprisingly the model can give some useful insights, and provides an initial description of electrical and thermal conductivity, the Hall effect and the Wiedemann–Franz law. Nevertheless, this classical model has no way of distinguishing between conductors and insulators and gives an incorrect prediction of the heat capacity of the electrons, so that ultimately a more comprehensive model is needed.

3.1 ELECTRONS AS CLASSICAL PARTICLES IN MATERIALS

What models can be developed for the behaviour of electrons in materials?

So far we have paid no attention to the effects which are due to the electrons in a solid. However, it was realized many years ago that, for example, the electrical conductivity of a metal was due exclusively to the motion of electrons inside the material.

The classical free-electron model was developed by Drude and Lorentz [1–4], in which it was assumed that electrons were classical particles with a kinetic energy $3k_B T/2$ where T is the thermodynamic temperature and k_B is Boltzmann's constant. These free electrons, which comprise only a small fraction of the total number of electrons in each atom, could account for both the electrical and thermal conductivity of a metal and also the optical reflectance at low frequencies.

Given the simplicity of this model, its successes are quite impressive. The most quoted success was the explanation of the Wiedemann–Franz law [5] by the model. One notable failure, however, was its prediction of the electronic contribution to the heat capacity. According to this model each electron should contribute $3k_B/2$ to the total heat capacity. In practice, this is not observed and in fact the electronic heat capacity is smaller by two orders of magnitude than expected on the basis of this theory.

Before going on to look at the model in detail let us consider how much of the volume inside a typical material is actually occupied by the atomic cores, and how much is 'empty space' so to speak. Consider for example, a material in which the lattice spacing is a and the radius of the ionic cores is r and which forms a body centred cubic structure. Since there are two atoms or ions per unit cell, the fraction of space occupied by the atoms or ions $8/3\pi(r/a)^3$. In sodium for example $r = 0.98 \times 10^{-10}$ m and $a = 4.2 \times 10^{-10}$ m. Therefore the fraction of space occupied by the atoms or ions is about 11%, meaning that the majority of the volume of the material is simply empty space for the electrons to move through.

3.1.1   Basis of the classical model

How might electrons behave inside a material?

The Drude model assumes that the metal behaves like an empty box containing a free-electron 'gas' (Fig. 3.1). The free-electron gas consists of the outer conduction electrons of the individual atoms. So a monovalent metal such as sodium contributes one electron per atom.

Therefore, the number of free electrons per unit volume will be,

$$N = \frac{N_0\rho}{W_A} Z$$

where Z is the valence of the atom, ρ is the density and W_A the atomic mass. These electrons have energies which are due to the thermal energy of the material and according to the model behave like the atoms of a gas in the kinetic theory

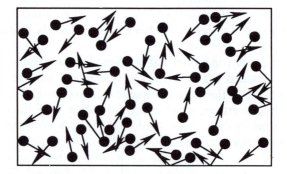

Fig. 3.1 Classical free-electron model of a solid consisting of a box containing a gas of free particles which obey the kinetic theory.

of gases. The number density of conduction electrons in a solid is typically 10^{28} m^{-3} (10^{22} cm^{-3}) which is of course about three orders of magnitude greater than for a typical gas at normal temperatures and pressures.

This should cause us some concern over the viability of the model. Another concern is of course that the material can hardly be considered to be an empty box since there are ions located on the lattice sites and these are electrically charged. Despite this, the Drude model rigidly follows the methods of the kinetic theory of a dilute, neutral gas, using Maxwell–Boltzmann statistics [6, 7].

The assumptions of the model are as follows.

1. Collisions between electrons are instantaneous and lead to scattering.
2. Between these collisions other interactions of the electrons with each other and with the ions are neglected in detail (although the interactions with the lattice are incorporated implicitly through an averaged resistive term in the equation of motion).
3. The mean free time of the electron between collisions is τ, and this time is independent of the electron's position and velocity.
4. Electrons achieve thermal equilibrium with their surroundings only through collisions with other electrons.

We will now develop the theory to help explain some of the well-known electrical and thermal properties of metals. According to the kinetic theory, the kinetic energy of each electron which is due to thermal energy will be $3k_B/2$. In the absence of an applied field the direction of motion is random.

3.2 ELECTRICAL PROPERTIES AND THE CLASSICAL FREE-ELECTRON MODEL

What effect does an applied electric field have on the electrons inside the material?

In the absence of a field, the electrons move randomly in all directions and consequently the net drift velocity is zero. When an electric field ξ is applied the electrons are accelerated with force $F = e\xi$, so that

$$m\frac{d\boldsymbol{v}}{dt} = e\xi$$

where m is the mass of the electrons (9.109×10^{-31} kg), e is the charge on the electrons (-1.602×10^{-19} C), and $\boldsymbol{v}$ is the velocity of the electrons.

There are also interactions between the drifting electrons and some of the lattice ions. This leads to resistance in the metal and ensures that the electrons are not accelerated indefinitely when a field is applied. Therefore an additional resistive term is needed. The equation of motion then becomes,

$$m\frac{d\boldsymbol{v}}{dt} = e\xi - \gamma\boldsymbol{v}$$

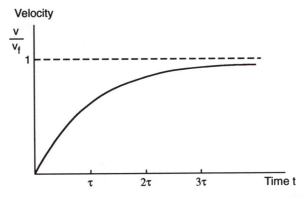

Fig. 3.2 Variation of the velocity of free electrons with time. The electrons reach a terminal velocity which is dependent on the resistive force caused by interactions with the lattice.

where γ is a constant which represents a resistive force which is proportional to electron velocity, preventing the electrons from being accelerated to infinite velocity.

The above equation is analogous to that encountered in viscosity, in which the resistive force is proportional to the velocity v. The electrons therefore reach an equilibrium velocity under the action of an accelerating field, and this equilibrium velocity is determined by the interactions with the lattice (Fig. 3.2).

3.2.1   Electrical conductivity

Can we obtain a 'first principles' expression for the electrical conductivity from the motion of the free electrons?

In the steady state, when the electrons have reached a final velocity v_f the force due to the electric field is equal to the resistive force, and opposite in direction,

$$m\frac{dv}{dt} = e\xi - \gamma v_f = 0$$

and so the coefficient γ can be obtained from

$$\gamma = \frac{e\xi}{v_f} = \frac{e}{\mu}$$

where v_f/ξ is known as the mobility of the electrons and is usually denoted μ. The velocity can be determined from the equation of motion,

$$m\frac{dv}{dt} + \frac{e\xi}{v_f}v = e\xi$$

which leads to the solution

$$v = v_f \left\{ 1 - \exp\left(-\frac{e\xi}{m v_f} t \right) \right\}.$$

We can define a relaxation time τ as,

$$\tau = \frac{m v_f}{e\xi} = \frac{m}{\gamma}$$

so that

$$v = v_f \left\{ 1 - \exp\frac{t}{\tau} \right\}.$$

The current density J is the product of the number of charge carriers per unit volume N, the charge per carrier e and the velocity v

$$J = Nev.$$

Since N should really be the number density of free electrons, rather than the total number of electrons per unit volume, we will use N_f instead of N,

$$J = N_f v_f e$$

and remembering that the final drift velocity is $v_f = e\xi\tau/m = e\xi/\gamma$

$$J = \frac{N_f e^2 \tau}{m} \xi$$

or, since $J = \sigma\xi$,

$$\sigma = \frac{N_f e^2 \tau}{m} = \frac{N_f e^2}{\gamma}.$$

Therefore we have derived a direct relationship between the electrical conductivity σ and the resistive coefficient γ from the equation of motion of the electrons. Typically for a metal τ is 10^{-14}–10^{-15} seconds, N_f is 10^{28} m^{-3}, e is 1.6×10^{-19} C and m is 9.1×10^{-31} kg, giving,

$$\sigma = 0.28 - 2.8 \times 10^6 \ (\Omega\text{m})^{-1}.$$

3.2.2 Ohm's law

Can we go even further and derive the well-known expression of Ohm's law from this free-electron model?

If we rearrange the above equations putting the current density $J = I/A$ where I is the current and A is the cross-sectional area, and the electric field $\xi = V/l$

where V is the voltage and l the length

$$\frac{I}{A} = \frac{N_f e^2 \tau}{m} \frac{V}{l}$$

$$I = \frac{N_f e^2 \tau A}{ml} V.$$

This then is an expression for Ohm's law which gives the relationship between current in an electrically conducting sample and the voltage across the sample,

$$I = \frac{V}{R}$$

so that

$$R = \frac{ml}{N_f e^2 \tau A} = \frac{l}{\sigma A}.$$

We can see, therefore, that the familiar macroscopic relationship known as Ohm's law [8] can be derived on the basis of the classical free-electron model. This gives some confidence in the model because it enables us to predict a known law on the basis of the model, and to relate model parameters to measurable properties.

3.3   THERMAL PROPERTIES AND THE CLASSICAL FREE-ELECTRON MODEL

How can the thermal properties be explained on the basis of the free-electron model?

We now look at the explanation of thermal conductivity and the Wiedemann–Franz law provided by the classical free-electron theory. In this the theory gives a satisfactory description.

3.3.1   Thermal conductivity

Can we obtain an expression for the thermal conductivity from the classical free-electron model equations?

The thermal conductivity K is the rate of transfer of thermal energy per unit thickness per unit area per unit temperature gradient, (see section 1.5.1),

$$K = -\frac{1}{A} \frac{dQ}{dt} \frac{dx}{dT}.$$

We can define $(dQ/dt)/A$ as the thermal current density, J_Q, so that in one

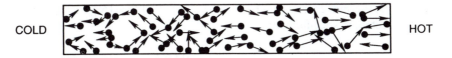

COLD   HOT

Fig. 3.3 Free electrons moving in a material under the action of a one-dimensional temperature gradient.

dimension

$$J_Q = -K \frac{dT}{dx}.$$

Consider a one-dimensional temperature gradient in a material which contains free electrons, as shown in Fig. 3.3.

Let $E(T)$ be the thermal energy per electron at temperature T, and the temperature at a point x be $T(x)$. An electron whose last collision was at another point x' will have an energy $E(T(x'))$ because the electrons can only gain or lose energy by colliding with other electrons, according to the theory. Assuming that the time between collisions is τ, those electrons from the high temperature end will have energy

$$E = E\{T(x - v\tau)\}.$$

Their contribution to the thermal current density will be the product of the number of electrons moving towards the cold end ($\leftarrow$), let us say $N_f/2$, their velocity v and their energy E

$$\Delta J_Q^{\leftarrow} = (N_f v/2)E\{T(x - v\tau)\}.$$

Similarly for electrons moving towards the hot end of the material ($\rightarrow$)

$$\Delta J_Q^{\rightarrow} = (N_f v/2)E\{T(x + v\tau)\}$$

the net contribution to the thermal current density is then

$$J_Q = (N_f v/2)[E\{T(x - v\tau)\} - E\{T(x + v\tau)\}]$$

and if the temperature difference over the range $v\tau$ is small, then

$$E\{T(x - v\tau)\} - E\{T(x + v\tau)\} = -2v\tau(dE/dx)$$

and therefore

$$J_Q = N_f v^2 \tau \frac{dE}{dT}\left(-\frac{dT}{dx}\right).$$

In fact to be exact, since we have a distribution of velocities, we should use the mean-square value of velocity for the electron gas, so the equation becomes,

$$J_Q = N_f \langle v^2 \rangle \tau \frac{dE}{dT}\left(-\frac{dT}{dx}\right)$$

where E is the thermal energy per electron, and $N_f(dE/dT) = (N_{Tot}/V)(dE/dT) = C_v^e$ is the electronic specific heat

$$J_Q = \langle v^2 \rangle \tau C_v^e \left(-\frac{dT}{dx} \right).$$

By comparison with the above equations, we see that we now have an expression for the thermal conductivity K,

$$K = \langle v^2 \rangle \tau C_v^e$$

or in three dimensions,

$$K = \tfrac{1}{3} \langle v^2 \rangle \tau C_v^e$$

where $\langle v^2 \rangle$ is the mean-square electronic velocity.

This shows that the free-electron theory is able to account for the thermal conductivity of a metal. Typical values of $\langle v^2 \rangle$, τ and the electronic specific heat C_v^e are

$$\langle v^2 \rangle = 5 \times 10^{10}\,\mathrm{m^2\,s^{-2}}$$
$$\tau = 2 \times 10^{-14}\,\mathrm{s}$$

and using,

$$N_f = 5 \times 10^{28}\,\mathrm{m^{-3}}$$
$$C_v^e = 1 \times 10^6\,\mathrm{J\,s^{-1}\,m^{-1}\,K^{-1}}$$

this gives the thermal conductivity as,

$$K = 3 \times 10^2\,\mathrm{J\,s^{-1}\,m^{-1}\,K^{-1}}$$

which compares well with the values of thermal conductivity K for copper, silver and gold which are 3.98×10^2, 4.29×10^2 and $3.15 \times 10^2\,\mathrm{J\,s^{-1}\,m^{-1}\,K^{-1}}$ respectively.

3.3.2   The Wiedemann–Franz law

Can the relationship between electric and thermal conductivities be explained by the classical free-electron model?

We have mentioned the Wiedemann–Franz law [5] in the previous chapter. The law can be explained on the basis of the classical free-electron model. Expressions have been obtained for both the electrical and thermal conductivities, and therefore the ratio can be expressed in terms of the electronic properties

$$\frac{K}{\sigma} = \frac{1}{3} \frac{\langle \mathbf{v}^2 \rangle C_\mathrm{v}^\mathrm{e} m \tau}{N_\mathrm{f} e^2 \tau}$$

$$= \frac{1}{3} \frac{m C_\mathrm{v}^\mathrm{e} \langle \mathbf{v}^2 \rangle}{N_\mathrm{f} e^2}$$

Using the relation between the heat capacity and the number of free electrons,

$$C_\mathrm{v}^\mathrm{e} = \tfrac{3}{2} N_\mathrm{f} k_\mathrm{B}$$

and relating the kinetic energy of the electrons to the thermal energy,

$$\tfrac{1}{2} m \langle \mathbf{v}^2 \rangle = \tfrac{3}{2} k_\mathrm{B} T,$$

this gives the following ratio between thermal and electrical conductivity,

$$\frac{K}{\sigma} = \frac{3}{2} \left(\frac{k_\mathrm{B}}{e} \right)^2 T$$

and therefore

$$\frac{K}{\sigma T} = \frac{3}{2} \left(\frac{k_\mathrm{B}}{e} \right)^2$$

which is a constant with value $1.11 \times 10^{-8} \, \mathrm{W\,\Omega\,K^{-1}}$ which is in reasonable agreement with the Wiedemann–Franz law, giving a value about half of that observed experimentally. The value of $3k_\mathrm{B}^2/2e^2$ is known as the Lorentz number.

So the theory does not quite predict the correct result. However, it does remarkably well for such a simple model and if this were the only problem, it could be overcome with some corrections. As we shall see there are more serious problems with the model which ultimately require a new approach to the description of electrons in metals.

In fact Drude's original calculation used $E = 3k_\mathrm{B} T$ as the energy per electron (assuming all thermal energy was kinetic energy) leading to a much better agreement with the Wiedemann–Franz law with a value of $K/\sigma T = 2.2 \times 10^{-8} \, \mathrm{W\,\Omega\,K^{-1}}$ which is exact to within experimental error. It was later shown by Lorentz [3] that the correct expression is $E = 3k_\mathrm{B} T/2$ because of the equipartition between potential and kinetic energy.

3.4 OPTICAL PROPERTIES OF METALS

How can the classical free-electron model explain the interaction of light with electrons in materials?

We have already given the equation of motion of free electrons in a solid as,

$$m \frac{\mathrm{d}\mathbf{v}}{\mathrm{d}t} + \gamma \mathbf{v} = \mathbf{F}$$

where F is the force on the electrons due to any stimulus. If we consider the excitation due to an incident light beam, then the electric field ξ of the incident light is,

$$\xi = \xi_0 \exp(i\omega t)$$

and since the force on a charge e is given by $F = e\xi$, the force on the electrons due to the incident light is

$$F = e\xi_0 \exp(i\omega t).$$

It has been shown above that the coefficient γ is equal to

$$\gamma = \frac{N_f e^2}{\sigma_0}$$

where N_f is the number of free electrons per unit volume, and where we have written σ_0 to distinguish the DC electrical conductivity instead of σ, and e is the electronic charge. Therefore the equation of motion of the electrons is

$$m\frac{d^2 x}{dt^2} + \frac{N_f e^2}{\sigma_0}\frac{dx}{dt} = e\xi_0 \exp(i\omega t)$$

We must expect a sinusoidal solution of this equation of motion for the electrons of the form $x = x_0 \exp(i\omega t)$. Inserting this into the differential equation yields the following expression for the amplitude of oscillation x_0 of the electrons

$$x_0 = \frac{\xi}{\left(\dfrac{N_f e\omega}{\sigma_0}\right)i - \dfrac{m\omega^2}{e}}$$

3.4.1   Dielectric polarization and absorption

How does the classical free-electron model account for the dielectric constant of a metal?

Since the electric polarization P is given by $P = eN_f x_0$, and the relative dielectric constant by $\varepsilon_r = 1 + (P/\varepsilon_0 \xi)$ the classical free-electron model necessarily leads to the following equation for the dielectric constant

$$\varepsilon_r = 1 + \frac{1}{\varepsilon_0}\frac{N_f e}{\dfrac{N_f e\omega}{\sigma_0}i - \dfrac{m\omega^2}{e}}$$

$$= 1 + \frac{1}{\dfrac{\varepsilon_0 \omega}{\sigma_0}i - \dfrac{\varepsilon_0 m\omega^2}{N_f e^2}}$$

If we then substitute for the two characteristic frequency terms,

$$\omega_1 = \sqrt{\frac{e^2 N_f}{m\varepsilon_0}} = 2\pi v_1$$

and,

$$\omega_2 = \frac{\varepsilon_0 \omega_1^2}{\sigma_0} = 2\pi v_2$$

this gives

$$\varepsilon_r = 1 + \frac{\omega_1^2}{i\omega\omega_2 - \omega^2} = 1 + \frac{v_1^2}{ivv_2 - v^2}.$$

v_1 is called the 'plasma frequency', and v_2 is called the 'damping frequency'. If we separate the dielectric constant into real and imaginary components, then on the basis of the classical free-electron model the real component of the relative dielectric constant, the polarization, is

$$\varepsilon_1 = n^2 - k^2 = 1 - \frac{v_1^2}{v^2 + v_2^2}$$

and the imaginary component, the absorption, is

$$\varepsilon_2 = 2nk = \frac{v_2}{v}\left(\frac{v_1^2}{v^2 + v_2^2}\right).$$

This shows that the two independent optical constants n and k (or alternatively ε_1 and ε_2) can be derived theoretically from the Drude free-electron model.

The variations of the polarization ε_1 and the absorption ε_2 with frequency according to the classical Drude free-electron model are shown in Figs. 3.4 and 3.5.

Table 3.1 Plasma and damping frequencies of various materials for the Drude free-electron model

Material	Plasma frequency $v_1\,(s^{-1})$	Damping frequency $v_2(s^{-1})$
Li	1.7×10^{15}	10.0×10^{12}
Na	1.4×10^{15}	4.8×10^{12}
K	1.0×10^{15}	3.1×10^{12}
Rb	0.9×10^{15}	4.8×10^{12}
Cs	0.75×10^{15}	5.15×10^{12}
Cu	......	4.7×10^{12}
Ag	......	4.4×10^{12}
Au	......	5.9×10^{12}

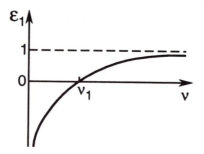

Fig. 3.4 Polarization ε_1 as a function of frequency according to the classical free-electron theory of metals.

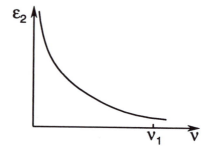

Fig. 3.5 Absorption ε_2 as a function of frequency on the basis of the classical free-electron theory of metals.

The 'plasma frequency' v_1 marks the boundary between the low-frequency reflecting regime and the high-frequency transparent regime.

3.4.2 Optical reflectance

Can the classical free-electron model explain the dependence of optical reflectance on wavelength of light in metals?

The variation of reflectance on the basis of the free-electron model can now be derived. The reflectance at normal incidence is related to the two optical constants by the relation,

$$R = \frac{(n-1)^2 + k^2}{(n+1)^2 + k^2}$$

and consequently

$$R = \frac{(\varepsilon_1^2 + \varepsilon_2^2)^{1/2} + 1 - \{2[(\varepsilon_1^2 + \varepsilon_2^2)^{1/2} + \varepsilon_1]\}^{1/2}}{(\varepsilon_1^2 + \varepsilon_2^2)^{1/2} + 1 + \{2[(\varepsilon_1^2 + \varepsilon_2^2)^{1/2} + \varepsilon_1]\}^{1/2}}$$

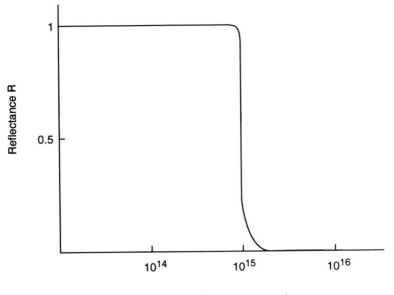

Fig. 3.6 Variation of reflectance with frequency of incident light according the free-electron model with $v_1 = 1 \times 10^{15}\,\mathrm{s}^{-1}$ and $v_2 = 1 \times 10^{12}\,\mathrm{s}^{-1}$.

when the above values of ε_1 and ε_2 can be substituted into the equation. The actual expression remains cumbersome, but the limits as $v \to 0$ and $v \to \infty$ can be found relatively easily, see Exercise 3.2. The varition of reflectance with frequency of incident light is shown in Fig. 3.6.

3.4.3   The Hagen–Rubens law

How can the well-known relationship between electrical conductivity and optical reflectance be explained?

From Maxwell's electromagnetic equations a relationship can be derived between the optical constants n and k, the dielectric coefficients ε_1 and ε_2, and the electrical conductivity σ. The relative dielectric constant ε_r is related to n and k by the equation,

$$\varepsilon_r = \varepsilon_1 + i\varepsilon_2 = n^2 - k^2 + 2nik$$

and the expression for ε_2 in terms of conductivity and frequency is, in SI units,

$$\varepsilon_2 = \frac{\sigma(\omega)}{\omega\varepsilon_0} = \frac{\sigma(v)}{2\pi v\varepsilon_0}$$

where v is the frequency of the electric field vector in the incident electromagnetic wave and ω is the angular frequency, $\omega = 2\pi v$.

In a metal the absorption ε_2 is much higher than the polarization ε_1, so that,

$$\varepsilon_r \simeq i\varepsilon_2$$

and the refractive index is related to the dielectric constants by

$$n^2 = \frac{\varepsilon_1}{2} + \frac{\sqrt{\varepsilon_1^2 + \varepsilon_2^2}}{2}$$

$$= \frac{\varepsilon_1}{2} + \frac{\sqrt{\varepsilon_1^2 + \sigma^2/\omega^2\varepsilon_0^2}}{2}.$$

Since $\sigma^2/\omega^2\varepsilon_0^2 \gg \varepsilon_1^2$, this gives

$$n^2 \simeq \frac{\sigma}{2\omega\varepsilon_0}.$$

Furthermore, since $\varepsilon_2 = 2nk = \sigma/\omega\varepsilon_0$, it is clear that we must also have

$$k^2 \approx \frac{\sigma}{2\omega\varepsilon_0} = n^2.$$

At longer wavelengths and hence lower frequencies v, the high conductivity σ leads to values of both k and n which are much greater than unity, and are typically of the order of 10^4.

Now considering the expression for the reflectance

$$R = \frac{(n-1)^2 + k^2}{(n+1)^2 + k^2}$$

$$= 1 - \frac{4n}{(n+1)^2 + k^2}$$

and since $n \approx k \gg 1$, this leads to

$$R \approx 1 - \frac{2}{n}$$

$$R \approx 1 - 2\sqrt{4\pi\varepsilon_0 v/\sigma}$$

$$= 1 - 2\sqrt{2\omega\varepsilon_0/\sigma}$$

which is the Hagen–Rubens relation [9], showing that high conductivity materials have high reflectance at long wavelengths. This provides the physical justification for our observation at the outset in Chapter 1 that good electrical conductors are also good optical reflectors.

3.4.4 Extensions of classical free-electron theory to optical properties at high frequencies

How can the higher frequency absorption bands be explained by the model?

At higher frequencies it is known from experimental observation that the reflectance does not necessarily remain low, but can show some localized peaks. These can be explained by an extension of the classical free-electron theory due to Lorentz [3, 4], in which some electrons behave as classical bound oscillators. These electrons are more tightly bound to the atoms and therefore can only respond to higher energy excitations. This leads to the following equation of motion for the electrons,

$$m\frac{d^2 x}{dt^2} + \gamma\frac{dx}{dt} + kx = e\xi_0 \exp(i\omega t)$$

where now the additional term kx represents a binding force between electrons and ionic sites. This equation describes the motion of these bound oscillators and gives absorption at higher frequencies in the form of bound oscillator resonances (Fig. 3.7).

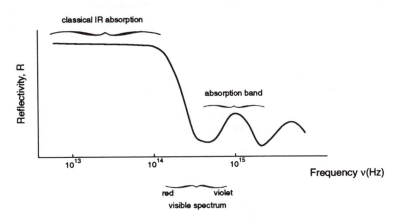

Fig. 3.7 Optical reflectance of metals beyond the infra-red range, in which 'resonances' at higher energies are observed. These can be attributed to bound oscillators rather than free electrons. Reproduced with permission from R. E. Hummel, *Electronic Properties of Materials* 2nd edn, published by Springer Verlag, 1993.

3.4.5 The photoelectric effect

What happens to the conduction electrons when high-energy light impinges on certain metals?

It was shown by Hertz [10] that a metallic surface emits electrons when illuminated by light of a very short wavelength. The emission of electrons from the surface is dependent on the wavelength of the light and not on the total energy incident on the surface.

The emission of electrons does not occur when the surface is irradiated with longer wavelength light over a longer time period, if the wavelength is below a certain critical value. In other words, if the frequency of the incident light is below a certain threshold value, exposure for longer periods will not lead to the emission of electrons, even though the total energy absorbed by the surface can be increased indefinitely in this way.

Furthermore, the kinetic energy E_K of the emitted electrons is dependent on the frequency of the light, but not on the intensity of the light,

$$E_K = \text{constant}(v - v_0)$$

where v is the frequency of incident light and v_0 is the threshold frequency which just enables electrons to escape from the material (Fig.3.8).

An explanation of these observations was given by Einstein [11]. If ω is the angular frequency of the incident radiation and $\hbar$ is Planck's constant divided by 2π, the energy of an incident light photon is

$$E(\omega) = \hbar\omega.$$

Now considering the electrons as classical particles trapped in a finite square-well potential of height ϕ, and assuming one light quantum interacts with one electron, the energy imparted contributes to the energy needed to overcome

Table 3.2 Values of the work function and threshold frequencies for the photoelectric effect in various materials

Material	Threshold energy or work function ϕ (eV)	Threshold frequency $v_0 (10^{14} s^{-1})$
Caesium	1.91	4.62
Rubidium	2.17	5.25
Potassium	2.24	5.42
Lithium	2.28	5.51
Sodium	2.46	5.95
Zinc	3.57	8.63
Copper	4.16	10.06
Tungsten	4.54	10.98
Silver	4.74	11.46
Platinum	6.30	15.23

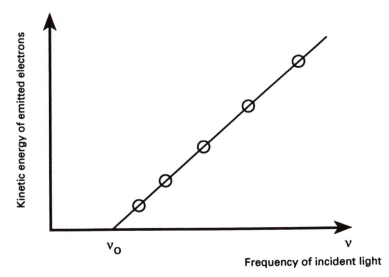

Fig. 3.8 Kinetic energy of emitted electrons in the photoelectric effect as a function of frequency of incident light.

the binding energy E_B of the electron to the solid and to the final kinetic energy of the electron,

$$\hbar\omega = E_K + E_B$$
$$= \tfrac{1}{2}m\boldsymbol{v}^2 + \phi$$
$$= \tfrac{1}{2}m\boldsymbol{v}^2 + \hbar\omega_0$$

where ϕ is the symbol used for the work function of the metal. This is identical to the threshold energy needed to liberate the electrons. Rearranging the equation gives,

$$E_K = h\nu - h\nu_0$$

which agrees with experimental observations. This model of the photoelectric effect we may call semi-classical. It relies on the quantum nature of the incident light, but still treats the electrons as classical particles in a finite potential well. The work function ϕ is the energy needed to extract one electron from the material, or alternatively is the depth of the potential well.

3.5 CONCLUSIONS

In the final analysis what are the advantages and disadvantages of the free-electron model?

The classical free-electron theory provides a simple model of the behaviour of electrons within a solid which seems to work reasonably well in certain cases for metals, for example the alkali metals, but not for insulators. The model takes no account of the properties of the lattice and conceptually this is a shortcoming. Nevertheless, there are a number of physical properties of solids which are dominated by the electrons rather than the lattice such as electrical and thermal conductivities.

The greatest success of the model was its ability to predict the relation between electrical and thermal conductivity known as the Wiedemann–Franz law. It also allows for a phenomenological explanation of electrical resistance and Ohm's law can be derived on this basis. Even the Hall effect can be described in the alkali and noble metals using this model. When combined with the quantum theory of light it gives an explanation of the photoelectric effect. The Drude model is also able to account for the optical properties of metals in the infra-red range and when combined with the Lorentz theory of bound oscillators it can account for the optical properties at higher frequencies.

However despite all these successes the simple Drude model has several shortcomings. The greatest of these arises in the calculation of the electronic specific heat capacity, and also its prediction of the optical properties of metals in the visible and ultra-violet ranges of the spectrum. It is unable to account for the temperature dependence of the DC conductivity (without the use of an *ad hoc* dependence of τ on temperature). Also it has not accounted for the temperature and field dependence of the Hall coefficient.

Consequently an improved theory of the electronic structure of materials is needed. This should show, first of all, that the heat capacity is 100 times smaller than is predicted by the classical theory. In order to address this problem it is quite clear that a simple modification of the classical particle model, such as replacing the 'empty box' of the solid by a periodic lattice with Coulomb-type interactions, will not be sufficient. Even though such an approach would, by its very nature, bring the model closer to reality, it would be unable to resolve the critical problem of heat capacity. Therefore a more radical approach is needed. This can be achieved by the inclusion of a quantum mechanical description of the free electrons.

REFERENCES

1. Drude, P. (1900) *Ann. der Physik*, **1**, 566.
2. Drude, P. (1900) *Ann. der Physik*, **3**, 369.
3. Lorentz, H. A. (1905) *Proc. Amst. Acad.*, **7**, 438.
4. Lorentz, H. A. (1909) *The Theory of Electrons*, Leipzig.
5. Wiedemann, G. and Franz, R. (1853) *Ann. Phys.*, **89**, 497.
6. Maxwell, J. C. (1860) *Phil. Mag.*, **19**, 19; 1860 **20**, 21; 1866 **32**, 390.
7. Boltzmann, L. (1868) *Sitz. Math–Naturwiss Cl. Akad. Wiss. Wien*, **58**, 517; 1872 **66**, 275; 1877 **75**, 67.
8. Ohm, G. S. (1827) *The Galvanic Circuit Investigated Mathematically*, T. H. Riemann, Berlin, reprinted by Krans Publishers, New York 1969.
9. Hagen, E. and Rubens, H. (1903) *Ann. Phys.*, **11**, 873.

10. Hertz, H. (1887) *Sitzungsberichte d. Berl. Akad. d. Wiss.*; 1887, *Wiedemann's Annalen*, **31**, 983.
11. Einstein, A. (1905) *Ann. der Physik*, **17**, 132.

FURTHER READING

Ashcroft, N. W. and Mermin, N. D. (1976) *Solid State Physics*, W. B. Saunders, Philadelphia, Chapter 1.
Chambers, R. G. (1990) *Electrons in Metals and Semiconductors*, Chapman and Hall, London, Chapter 1.
Dugdale, J. S. (1977) *The Electrical Properties of Metals and Alloys*, Edward Arnold, London, Chapter 2.

EXERCISES

Exercise 3.1   Drude free-electron theory of metals. Outline the basic assumptions of the classical (Drude) free-electron theory of metals. Describe the extension by Lorentz of the original theory to include the effects of bound electrons and describe how these were represented in the model.

Exercise 3.2   Reflectivity based on Drude theory. State the relationship between the dielectric constants of a material and the Drude parameters v_1 and v_2 from the classical free-electron model. Using the relationship between the reflectance R and ε_1 and ε_2 show the limiting values of R as $v \rightarrow 0$ and $v \rightarrow \infty$. Under which frequency conditions does the Drude prediction of R work and under which conditions is it inadequate?

Exercise 3.3   Electrical and optical properties of a classical free-electron metal. Write a short discussion of the principal successes and failures of the classical free-electron theory. The mobility of electrons is defined as the ratio of electric field to velocity ($\mu = v/\xi$). In a piece of copper the mobility was found to be $3.5 \times 10^{-3} \, \text{m}^2 \, \text{V}^{-1} \, \text{s}^{-1}$. Assuming that the classical free-electron model can be used, calculate the resistive coefficient γ, the electrical conductivity σ and the mean free time between collisions. (Copper has a density of $8940 \, \text{kg} \, \text{m}^{-3}$, an atomic weight of 63 and each atom donates one conduction electron.)

4

Conduction Electrons in Materials – Quantum Corrections

In this chapter we look at another 'free' electron model in which the electrons are described by wave functions contained within a material boundary. Once again the model assumes that the potential inside the material is completely flat, which amounts to ignoring the presence of the lattice. The electrons are therefore only constrained by the limits of the material. The main difference between this and the classical free-electron model is that only certain energy levels are allowed. This means that with constraints on the numbers of electrons which can occupy these energy levels, only those electrons with energies close to the top of the electron 'sea' can contribute to the heat capacity. This results in a reduction by two orders of magnitude of the expected electronic contribution to the heat capacity, bringing it into agreement with measured values. However, ultimately the model is insufficient because of its failure to take the lattice potential into account.

4.1 ELECTRONIC CONTRIBUTION TO SPECIFIC HEAT

How can the free-electron model be developed further to allow prediction of a wider range of properties?

In view of the apparent failures of the classical free-electron model of metals, we must look at the next level of sophistication to see whether an improved theory can be derived. The next procedure is to try a quantum mechanical approach to the free-electron model. The most critical problem in the Drude theory was the contribution to the heat capacity from the electrons. The theory predicted a contribution of $\frac{3}{2}k_B$ per electron, whereas in practice this contribution was known to be much smaller. The classically-expected electronic contribution to the specific heat is

$$C_v^e = \frac{3}{2} N_0 k_B.$$

With $N_0 = 6.02 \times 10^{26}$ atoms per kilogram mole, and assuming one conduction electron per atom

$$C_v^e = 12.5 \, \text{J} \, \text{mol}^{-1} \, \text{K}^{-1}.$$

The observed value is typically

$$C_v^e = 0.2 \, \text{J} \, \text{mol}^{-1} \, \text{K}^{-1}.$$

The discrepancy arises from the treatment of the conduction electrons as a classical free gas. The fact that $N_f = 10^{28}$ per cubic metre, which is 1000 times greater than for a classical free-electron gas, and also the fact that the particles are not electrically neutral (and therefore interact through Coulomb repulsion) should immediately have caused suspicion about the adoption of the classical kinetic theory for the model. It was found that the whole concept of classical statistical thermodynamics was inapplicable to this situation and that a new form of statistics was needed to describe electrons in solids.

4.2 WAVE EQUATION FOR FREE ELECTRONS

What is the equation of motion of the electrons in this case?

If quantum mechanical principles are used to describe the electrons the general expression for the time-independent Schrödinger equation [1] is

$$-\frac{\hbar^2}{2m} \nabla^2 \psi(x) + V(x)\psi(x) = E\psi(x)$$

where in our case m is the mass of an electron, $\hbar$ is Planck's constant divided by 2π and $V(x)$ is the potential energy at the point in space defined by the vector x. E is the energy of the electron, which we may consider as a wave or a particle. Note that the Schrödinger equation here is simply an expression of the conservation of energy. The sum of kinetic energy and potential energy on the left-hand side equals the total energy on the right-hand side.

We can find solutions of the wave equation in particular cases. Generally, the solution has the form

$$\psi(x) = A_1 \cos kx + A_2 \sin kx$$

in one dimension, where x is the spatial dimension, and k the wave vector may be real or imaginary. The probability of finding the electron at any given point x in space is

$$P(x) = \psi^*(x)\psi(x) = |\psi(x)|^2$$

where ψ^* is the complex conjugate of the wave function ψ.

The intention in this book is to solve the Schrödinger equation in one dimension only. This gives the greatest clarity of explanation by presenting the essential concepts behind the quantum mechanical free-electron model. Nothing new is learned by generalizing to three dimensions at this stage, other than additional detail.

4.2.1 Consequences of the quantum theory of free electrons

How does the energy of a free electron depend on its wave vector?

It is found, by substituting the solution of the wave function into the Schrödinger equation, that when the electrons are completely free, that is providing that $V(x)$ is everywhere zero, the energy E and the wave vector k are related by

$$k = \sqrt{\frac{2mE}{\hbar^2}}.$$

We can rearrange this so that

$$E = \frac{\hbar^2 k^2}{2m}.$$

The relationship between energy E and wave vector k is depicted by the free-electron parabola show in Fig. 4.1. This applies only to completely free electrons.

The important point here is that since the electrons are completely free they can be considered to be in the presence of a completely flat potential, $V = 0$ over all space. Therefore the electron wave function ψ can extend to infinity and there are no boundary conditions to be applied to the wave function.

With no further constraints to be applied, we see that all values of k are allowed solutions of the Schrödinger wave equation, and therefore all energies are allowed for free electrons. In other words, the allowed energy values E form a continuum for free electrons.

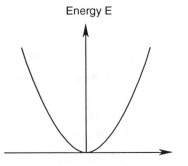

Fig. 4.1 Free-electron parabola showing the dependence of energy E on wave vector k in which all energies are possible.

If we extend the above one-dimensional model to three dimensions the components of the wave vector along the x, y and z directions are k_x, k_y and k_z. Consequently the energy of a free electron becomes

$$E = \frac{\hbar^2}{2m}(k_x^2 + k_y^2 + k_z^2).$$

This means that if the electrons are completely free, all electron states with the same energy form a sphere in k-space. Therefore if we have free electrons occupying all energy levels up to a maximum energy E_F, the electrons are all contained in a sphere in k-space. We shall return to this idea later in discussing the Fermi surface for free electrons.

4.3   BOUNDARY CONDITIONS: THE SOMMERFELD MODEL

How can a quantum mechanical description of the electrons be used while preserving most of the classical description from the previous model?

The first application of quantum mechanics to the problem of the electronic properties of metals was made by Sommerfeld [2, 3]. The initial motivation was to resolve some of the discrepancies between the classical Drude free-electron model and experimental observation, in particular those problems which arose over the electronic contribution to the specific heat.

The Sommerfeld model is still very much a free-electron model. It assumes that the conduction electrons reside within a potential which is constant everywhere inside the metal. This of course is only an approximation, but it is a useful first step because it leads to a relatively simple formulation of the problem.

Sommerfeld applied the Pauli exclusion principle [4] (see section 4.4.1) to the free-electron model of metals. This resulted in the resolution of the most serious anomalies in the classical electron theory of metals. In particular, the difficulty over the electronic contribution to the specific heat capacity was explained.

In its simplest form, the Sommerfeld model involves only this single modification to Drude theory. Later we shall look at further modifications such as the solution of the Schrödinger equation with more realistic assumptions such as periodic potentials. For the time being, however, we will just look at the solid as a simple, flat-bottomed potential box containing electrons. In this the solid is represented as a square-well potential. In the most elementary calculation an infinite square well is assumed, but a finite square-well potential also gives relatively simple solutions.

We will find that there are only certain allowed or accessible energy states for the electrons under these conditions. Note here that it is the boundary conditions that are crucial in determining the allowed energy states, not just the wave equation itself.

4.3.1   Wave equation for bound electrons in an infinite square-well potential

What happens to the allowed electronic states, that is to say solutions of the wave equation, if boundary conditions are imposed?

Let us now go to the other extreme and consider bound electrons. This would seem to be a quite reasonable approach to the problem of electrons in a solid since mostly the electrons are constrained to remain within the solid. The simplest possible model, then, is to suppose that the electrons move freely within the confines of the volume of the solid, but encounter an infinite potential at the boundary of the solid which prevents them from leaving the solid. We may represent this under the simplest conditions in one dimension as an infinite square-well potential as shown in Fig. 4.2.

The flat potential inside, which is constant everywhere, may seem like a gross approximation to a solid, but remember that most of the solid is 'empty space' anyway and the outer electrons are screened from the ionic cores by the localized electrons. If we allow the potential outside the range $-a \leqslant x \leqslant a$ to go to infinity, then the probability of the electron appearing in the ranges $x < -a$ or $x > a$ is zero. This means that,

$$\psi^*(x)\psi(x) = 0$$

outside the solid. Our boundary conditions therefore must be

$$\psi(x = a) = 0$$

$$\psi(x = -a) = 0.$$

If we then solve the one-dimensional wave equation in the time-independent case,

$$-\frac{\hbar^2}{2m}\frac{d^2\psi}{dx^2} + V(x)\psi = E\psi$$

we again obtain the solutions

$$\psi(x) = A_1 \cos kx + A_2 \sin kx.$$

The difference from the previous example of the completely free electron, is that now we need to apply the known boundary conditions. It is these boundary conditions which determine the allowed solutions.
At $x = a, \psi(a) = 0$, and so

$$\psi(a) = A_1 \cos ka + A_2 \sin ka = 0.$$

At $x = -a, \psi(-a) = 0$ and so,

$$\psi(-a) = A_1 \cos(-ka) + A_2 \sin(-ka) = 0$$

which leads to the solution, $A_1 = 0$ and,

$$ka = n\pi$$

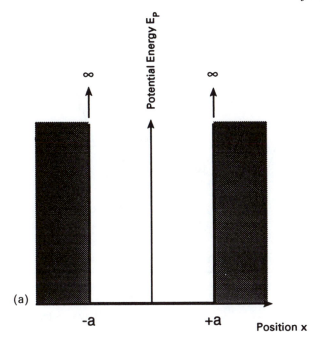

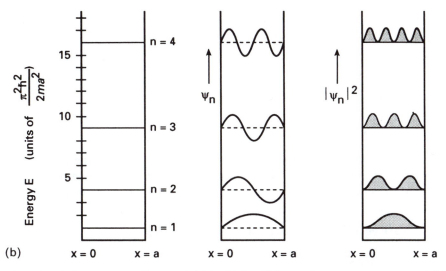

Fig. 4.2 Infinite square-well potential (a) which will be used to represent a material in which electrons are contained only by the physical limits of the material; (b) shows the solutions of the wave equation inside an infinite square-well potential. Notice that these produce standing waves in ψ.

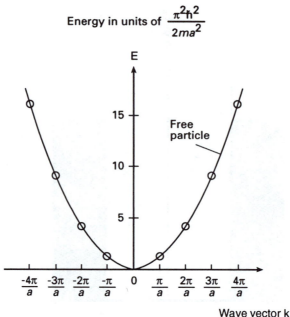

Fig. 4.3 The dependence of energy on wave vector for electrons confined in an infinite square-well potential. The parabolic relation still holds, but only certain discrete energy levels are allowed.

so that

$$\psi_n(x) = A_2 \sin\left(\frac{n\pi}{a}x\right).$$

This is the solution of the wave equation under the given boundary conditions. Note that only integer values of n satisfy the boundary conditions and so the allowed energy levels are now discrete (Fig. 4.3)

$$E_n = \frac{\hbar^2}{2m}\frac{n^2\pi^2}{a^2}   n = 0, 1, 2, 3, \ldots.$$

This means that only wave functions with certain isolated energies can fit into the solid and still meet the boundary conditions. The extension of this model to three dimensions is simple and introduces no new concepts.

4.3.2 Wave equation for electrons in a finite potential well

What are the allowed electronic states, i.e. solutions of the wave equation, when the potential box is finite?

The next level of complexity, which brings us closer to reality, is to consider the possible electron states in the presence of a finite potential well. Clearly, in

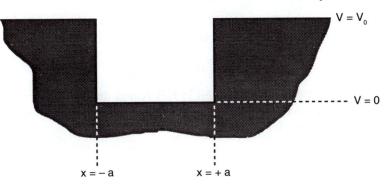

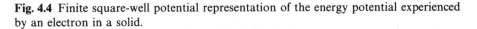

Fig. 4.4 Finite square-well potential representation of the energy potential experienced by an electron in a solid.

a real material the potential barriers marking the end of the solid will be finite. This now brings us close to the idea of the original Drude free-electron model which was a box, consisting of a finite potential barrier represented by the boundary of the solid, containing classical particles which we called electrons (Fig. 4.4). Now we will use the same idealized box to represent the solid boundary but instead fill it with wave-like representations of electrons instead of classical particles. The depth of the potential well must, of course, be finite because of the evidence supplied by the photoelectric effect and thermionic emission. In fact these two phenomena can be used to determine the depth V_0 of the potential well, or more correctly, the depth of the most energetic electrons below the top of the potential well.

Again we solve the Schrödinger equation. Inside the solid we assign for simplicity $V = 0$. The wave equation is then,

$$-\frac{\hbar^2}{2m}\frac{\mathrm{d}^2\psi}{\mathrm{d}x^2} = E\psi$$

where E is the energy of the electron.

Solutions of the wave equation for positions inside the box, that is for $|x| < a$, give the following wave function,

$$\psi = A_1 \cos kx + A_2 \sin kx$$

where $k = (2mE)^{1/2}/\hbar = 2\pi/\lambda$.

For lower energies $(E < V_0)$ λ is the wavelength of the standing waves represented by the wave function inside the potential box. Note that larger energies correspond to shorter wavelengths and vice versa. For higher energies $(E > V_0)$ the wave function will extend well beyond the boundaries of the box and therefore does not correspond to a standing wave even in the region of the potential box.

Outside the material, that is for $|x| > a$, we now have a finite potential $V = V_0$, so in this region the wave equation is,

$$\frac{d^2\psi}{dx^2} + \frac{2m}{\hbar^2}(E - V_0)\psi = 0.$$

The solutions of the wave equation in this region depend on the energy E. If $E < V_0$ then the second term on the left-hand side of the equation becomes negative and solutions for $|x| > a$ have the form,

$$\psi(x) = B_1 e^{k_1 x} + B_2 e^{-k_1 x}$$

where $k_1^2 = -2m(E - V_0)/\hbar^2$.

These solutions for $E < V_0$ in the region outside the box are not periodic. They represent an exponentially decaying waveform which means that for $E < V_0$ the wave function decays with distance beyond the limits of the material.

It is easy to see that since the wave function cannot diverge outside the potential well, we must have $B_1 = 0$ for $x > a$ and $B_2 = 0$ for $x < -a$ so we are left with

$$\psi(x) = B_2 e^{-k_1 x} \quad \text{for } x > a \text{ and } E < V_0$$

$$\psi(x) = B_1 e^{k_1 x} \quad \text{for } x < -a \text{ and } E < V_0.$$

In order to find the coefficients in those cases with $E < V_0$ we must again use the boundary conditions at $x = -a$ and $x = a$. Since we already have the form of solution above for both inside and outside the box, we merely need to ensure that these match at the boundaries of the box.

We will now proceed using only the antisymmetric component of the wave function, $\sin kx$. A similar argument can be applied to the symmetric wave function, $\cos kx$. The amplitude of ψ must be continuous across the boundary, so that, for the odd parity (sine) solutions,

at $x = a$,

$$A_2 \sin ka = B_2 e^{-k_1 a}$$

at $x = -a$,

$$A_2 \sin(-ka) = B_1 e^{-k_1 a}$$

and the derivative $d\Psi/dx$ must also be continuous across the boundary

at $x = a$,

$$A_2 k \cos ka = -k_1 B_2 e^{-k_1 a}$$

at $x = -a$

$$-A_2 k \cos(-ka) = k_1 B_1 e^{-k_1 a}.$$

We therefore simply equate the solutions both inside and outside the box at the boundaries and these lead to the condition

$$k \cot ka = -k_1.$$

Similarly, for the even parity (symmetric or cosine) solutions it can be shown by the above argument that

$$k \tan ka = k_1.$$

These last two equations may be solved graphically or numerically for the allowed energies E_n, remembering that $E = \hbar^2 k^2/2m$.

The important result here is that for energies $E < V_0$, which represent electrons 'contained' within the potential well, the allowed energies lead to a discrete set of energy levels. We say 'contained' because it is also clear from the wave function that the electron has a finite probability of being just outside the potential well, that is $\psi(x) \neq 0$ for $|x| > a$ (or more precisely $|\psi(x)|^2 \neq 0$ in this range, since the probability of observing the electron is dependent on $|\psi|^2$ rather than ψ).

This is a new result emerging from the quantum mechanical treatment of the problem, that does not arise in the classical description of the electrons in materials.

For energies $E > V_0$ the electron wave functions extend outside the limits of the potential well,

$$\psi(x) = B_1 e^{k_1 x} + B_2 e^{-k_1 x}$$

where $k_1^2 = -2m(E - V_0)/\hbar^2$, which gives an imaginary value of k_1, and hence the solutions are periodic for $E > V_0$.

The electrons with energies greater than V_0, the depth of the potential, are not constrained by the presence of the potential and can therefore have a continuous spectrum of energies and extend spatially beyond the box. We should note however that the wave functions of these higher energy electrons in the locations $|x| < a$ are perturbed by the presence of the potential well.

4.4 DISTRIBUTION OF ELECTRONS AMONG ALLOWED ENERGY LEVELS

How does quantum mechanics affect the distribution of electron energies?

Unlike classical statistical mechanics [5, 6] which allows any energy state to be occupied by any number of electrons, quantum mechanics imposes restrictions on the number of electrons which can occupy a given energy level (Fig. 4.5).

4.4.1 The Pauli exclusion principle

How many electrons can we fit into a given energy state?

The Pauli principle [4] states that no two electrons can have the same set of quantum numbers, and therefore cannot occupy identically the same energy level in our solid. We are ignoring electron spin for the moment. This leads to a radical change in our understanding of the ground energy state of a solid

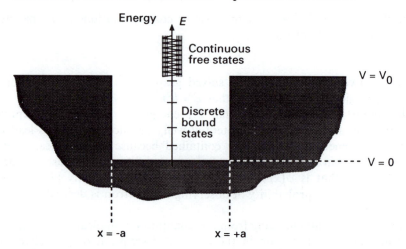

Fig. 4.5 Allowed energy states in a finite square-well potential obtained by a solution of the Schrödinger equation.

containing a number of electrons – since no longer can all of the electrons reach the nominal lowest energy level that they would be allowed to occupy in classical particle physics.

If we have a certain number N electrons in the solid, then we must begin by filling the lowest energy level with an electron. When this is filled we move up to the next energy level and so on. We do this until each electron has been assigned to the lowest remaining energy state available. This then is the ground state of our solid and corresponds to the electronic occupancy at absolute zero of temperature.

If we proceed in this way until all electrons have been assigned to an available energy state, there will exist a highest occupied energy level. This is known as the Fermi level. It separates the occupied from the unoccupied states only at 0 K (i.e. in the ground state of the solid). When we consider electrons occupying

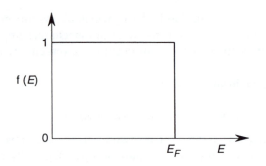

Fig. 4.6 Probability of occupancy of energy states in the ground state. All levels up to E_F are occupied, above that the energy levels are unoccupied.

available energy states, we can define a function $f(E)$ which describes the probability that a given energy state is occupied.

Clearly, because of the constraints of a limited discrete set of allowed energies, and the limitation of one electron per energy state (assuming we treat the spin up and spin down as distinct states), the probability of the lowest energy state being occupied is 1 and providing we have enough electrons available, then, at least in the ground state 0 K, the probability of any state being occupied remains 1 until we literally run out of electrons to occupy the allowed states. The probability of occupancy $f(E)$ at 0 K therefore must have the form, shown in Fig. 4.6, where E_F is the highest occupied level which is called the Fermi level.

4.4.2 The failure of classical Maxwell–Boltzmann statistics

Can Maxwell–Boltzmann statistics apply once we have restricted the number of allowed states for an electron to occupy?

In classical statistical mechanics, all energies are available and any number of particles can have an identical energy E, or equivalently have the same state. Under these conditions, the probability of a particle having an energy E at temperature T is

$$P(E) = P_0 \exp\left(-\frac{E}{k_B T} \right).$$

The normalization constant P_0 is found simply from the condition that the integral of the probabilities over all possible states must be unity

$$\int_{\substack{\text{all} \\ E}} P(E)\,dE = 1.$$

The Maxwell–Boltzmann distribution function has the form illustrated in Fig. 4.7.

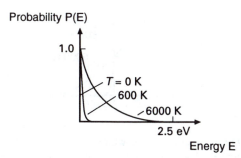

Fig. 4.7 The Maxwell–Boltzman distribution function which represents the probability of a classical particle being found with energy E.

Once we have accepted the idea of a finite number of discrete energy levels and applied the Pauli exclusion principle which limits occupancy of these available states, it is immediately apparent that classical Maxwell–Boltzmann statistics can no longer apply, since classical statistics is based on the concept that any energy can be taken by any number of particles.

4.4.3 Probability of occupancy

With a finite number of allowed electron states in a material, how do the electrons arrange themselves?

Since the electrons can no longer follow classical continuum statistics a new description of the energy spectrum of electrons is needed. The Fermi–Dirac function $f(E)$ which describes the probability of occupancy of electrons as a function of energy E [7, 8] has the form

$$f(E) = \frac{1}{1 + \exp\left(\dfrac{E - E_F}{k_B T}\right)}.$$

When $E = E_F$ it can easily be seen that $f(E) = 0.5$, and this corresponds to the Fermi energy, the highest occupied energy level when the electron configuration is in its ground state.

At temperatures above $0\,K$ the step function form of this probability distribution becomes distorted with a range of intermediate energies close to E_F for which $f(E)$ is neither 1 nor 0 (Fig. 4.8). The reason for this is that the thermal energy can excite electrons from some of the lower energy states to higher energy states. Therefore some lower energy states that were occupied

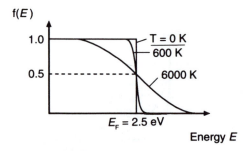

Fig. 4.8 The Fermi–Dirac function $f(E)$, which gives the probability of occupancy, varies with temperature. At absolute zero ($0\,K$), only the lowest energy states up to the Fermi level are occupied. At higher temperatures energy states above the Fermi level can be occupied.

at 0 K are now unoccupied and some higher energy states that were unoccupied at 0 K are now occupied.

4.4.4 Fermi–Dirac statistics

What statistics do the electrons obey?

When Drude was developing the classical free-electron model of a metal, it was natural to assume that the electron energy distribution was like that of a classical gas and obeyed Maxwell–Boltzmann statistics because no other statistical method for describing assemblies of particles was available at the time. This gave the number of electrons per unit volume with velocities in the range $v \pm dv$ as

$$f_{\mathrm{B}}(v) = N \left(\frac{m}{2\pi k_{\mathrm{B}} T} \right)^{3/2} \exp \left(-\frac{mv^2}{2k_{\mathrm{B}} T} \right).$$

However, since electrons behave according to the Pauli exclusion principle, the distribution of electrons with velocities in the range of $v \pm dv$ must be described by Fermi–Dirac rather than Maxwell–Boltzmann statistics. In this case the expression according to Fermi–Dirac statistics becomes

$$f(v) = \frac{1}{4} \frac{m^3}{h^3} \frac{1}{\exp((\frac{1}{2}mv^2 - k_{\mathrm{B}} T_0)/k_{\mathrm{B}} T) + 1}.$$

where $k_{\mathrm{B}} T_0$ is a normalization factor which ensures that the integral of the probability over all velocities is unity. It is also known as the chemical potential.

4.4.5 Electron energy distributions

What other information is needed to describe the electron distribution?

We can describe the number of available electron states as a function of energy E by the density of available states $D(E)$. This density of states is independent of the electrons available to fill the states, it is simply an expression of what energy values are allowed.

The density of occupied states $N(E)$ describes the number of electrons per unit energy interval as a function of energy. This is related to the density of available states $D(E)$ through the probability of occupancy $f(E)$ by the equation

$$N(E) = 2f(E) D(E)$$

where the factor 2 arises because electrons can have spin up and spin down [9, 10], and therefore each available energy state can be occupied by two electrons, one with spin up, the other with spin down. This simple expression allows the electron distribution to be described in terms of the available levels $D(E)$ as determined by the ionic potential, and the distribution of electrons among these levels $f(E)$ as determined by temperature.

4.4.6   Density of states

How does the number of allowed energy states vary with energy level?

We have talked at length about the allowed energy levels, how they are discrete states at low levels near to the atomic cores, and how they form almost continuous bands of allowed energies at higher levels. We have also shown that the electrons occupy these states beginning at the lowest energy level and continuing upwards to the Fermi level. In doing so they obey the Pauli exclusion principle. So far, however, we have made no mention of exactly how many allowed levels $D(E)$ there are for electrons at any given energy level or how this can be determined. We must, therefore, address this question of an equation for the density of states, at least in some simplified cases.

4.4.7   Model density of states in square-well potential

How does the density of states vary with energy in the simplest case of a three-dimensional square-well potential?

Returning to our familiar model, the square-well potential, we can investigate mathematically the density of available states. If we generalize the earlier result for the infinite square-well potential (section 4.3.1) to three dimensions, the energy of a given state is

$$E = \hbar^2 k^2/2m$$

$$E_n = \frac{\hbar^2 \pi^2}{2ma^2}(n_x^2 + n_y^2 + n_z^2)$$

where n_x, n_y, n_z are quantum numbers representing allowed solutions of the wave equation, such that

$$n^2 = n_x^2 + n_y^2 + n_z^2.$$

A given energy state, which represents a particular combination of n_x, n_y and n_z, can be represented as a point in quantum number space. All states with equal energy will lie on the surface of a sphere. Since only positive values of n_x, n_y and n_z are allowed, the available states are restricted to 1/8 of the volume of a sphere in quantum number space.

The number of distinct allowed states $N_0(E_n)$ with energies equal to or less than an energy E_n is

$$N_0(E_n) = \int_0^{E_n} D(E)\,\mathrm{d}E.$$

This is equal to the total volume of quantum number space divided by the volume occupied by one quantum state. Since each quantum state occupies unit volume in quantum number space this gives

$$N_0(E_n) = \frac{1}{8} \cdot \frac{4}{3} \pi n^3.$$

Using the free electron approximation, and remembering that there can be two electrons per energy state

$$2N_0(E_n) = \frac{\pi}{3} \left(\frac{2ma^2}{\pi^2 \hbar^2} \right)^{3/2} E^{3/2}.$$

We will use the $N_0(E)$ notation to remind us that it represents the number of states from $E = 0$ to $E = E$. If we differentiate this expression, we obtain the number of energy states between E and $E + dE$, which we will denote $D(E)$

$$D(E) = \frac{d}{dE}(N_0(E)) = \frac{\pi}{4} \left(\frac{2ma^2}{\pi^2 \hbar^2} \right)^{3/2} E^{1/2}.$$

Putting $V = a^3$, the volume of a cube of material of side a gives,

$$D(E) = \frac{V}{4\pi^2} \left(\frac{2m}{\hbar^2} \right)^{3/2} E^{1/2}$$

and since we can have two electrons per energy state, one with spin up and the other with spin down the actual number density of electron states is twice $D(E)$.

Remember, this is an approximate expression which is only valid for free electrons. In a real material the dependence of the density of states on the energy is modified and therefore not truly parabolic.

In order to describe adequately the allowed electron states in a solid we not only need the available energies, as determined by the boundary conditions imposed by the lattice, but also the density of states at each of the allowed

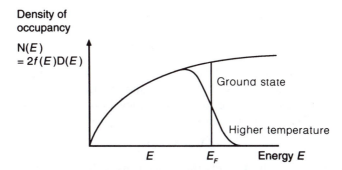

Fig. 4.9 Density of states as a function of energy E from quantum free-electron theory. Only those electrons within $k_B T$ of the top of the distribution can absorb thermal energy. This reduces the expected electronic heat capacity compared with classical theory.

energies. Generally the number of distinct available states $D(E)$ increases with energy E as shown in Fig. 4.9.

4.5   MATERIAL PROPERTIES PREDICTED BY THE QUANTUM FREE-ELECTRON MODEL

Do the predictions of the quantum free-electron model correlate with observations?

In this section we look at how the quantum free-electron model overcomes some of the failures of the classical free-electron model. In particular how it explains the low electronic contribution to the heat capacity, the temperature-independent paramagnetic susceptibility in metals, and thermionic emission.

4.5.1   Heat capacity

How does the electron heat capacity as predicted by the quantum free-electron model vary from that predicted by the classical free-electron model?

We now address directly the main question which proved the downfall of classical Drude theory – the discrepancy between the observed electronic heat capacity and the predicted value based on classical theory. The application of the Pauli principle and quantum mechanics to the problem of electrons in a solid leads to a new distribution of energy states given by Fermi–Dirac statistics.

When a solid is heated from absolute zero (0 K) not every electron is able to gain an energy $k_B T$ because the energy state which it would need to occupy in order to absorb the energy is either already occupied by another electron, or in some cases does not exist. Only those electrons in energy levels within $k_B T$ of the Fermi level can be thermally excited to higher available energy levels. This leads to a change in the density of occupied states as shown in Fig. 4.9. The result is a blurring of the distinction between occupied and unoccupied states, which at 0 K was very sharp.

The kinetic energy dE_K of the electrons above the Fermi level due to thermal energy is,

$$dE_K = \tfrac{3}{2}k_B T\, dN$$

where dN is simply the number of electrons above the Fermi level. This depends on the density of states at the Fermi level and the thermal energy. Assuming that the density of states close to the Fermi level does not change drastically as a function of energy E we can write,

$$dN = N(E_F)k_B T$$

because as shown in Fig. 4.9 only those electrons within $k_B T$ of the Fermi level can absorb thermal energy and contribute to the specific heat

$$dE_K = \tfrac{3}{2}k_B T\, N(E_F)k_B T.$$

Immediately we have a solution to the problem of the small electronic contribution to the heat capacity. If the total number of electrons is N then only a fraction of these can be thermally excited at a temperature T. This fraction is typically T/T_F where T_F is the temperature corresponding to the Fermi energy $k_B T_F = E_F$. The electronic heat capacity is then

$$C_v^e = \frac{dE}{dT}$$

$$= 3 k_B^2 T N(E_F).$$

For free electrons the density of occupied states at the Fermi level is given by,

$$N(E_F) = \frac{3}{2}\left(\frac{N}{E_F}\right)$$

so that

$$C_v^e = \frac{9}{2} N k_B \frac{T}{T_F}.$$

If we take quantum mechanics fully into account then the equation for the heat capacity is slightly modified to

$$C_v^e = \frac{\pi^2}{2} N k_B \frac{T}{T_F} = \frac{\pi^2}{2} \frac{N k_B^2 T}{E_F}$$

where N is the total number of conduction electrons per unit volume. At room temperature this value of C_v^e is about 1% of the classically expected electronic heat capacity of $3N_0 k_B/2$, because typically $T_F = 30\,000$ K, and hence $T/T_F = 0.01$.

4.5.2 Pauli spin paramagnetism

What predictions does the model make about other bulk properties, such as magnetic susceptibility?

Electrons have a magnetic moment associated with both their spin and orbital angular momentum. All metals exhibit a weak paramagnetism which is independent of temperature and this may be explained with the free-electron model of metals developed in this chapter [11]. When an external magnetic field is applied to a material the orientations of the magnetic moments of the electrons are constrained to lie either parallel or antiparallel to the field. This leads to a splitting of the energy of the parallel and antiparallel states as shown in Fig. 4.10.

Those electrons with magnetic moments m parallel to the field direction have energies reduced by $\Delta E = -\mu_0 mH$, while those with magnetic moments antiparallel to the field have energies increased by $\Delta E = \mu_0 mH$. Some of these

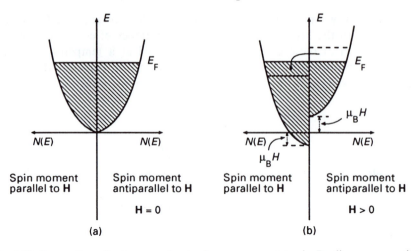

Fig. 4.10 Separation of spin-up and spin-down energy states in Pauli paramagnetism.

antiparallel electrons can reduce the energy of the system by occupying parallel states of lower energy.

The number of electrons which can change orientation and still reduce the total energy are those which were within $\mu_0 mH$ of the Fermi level in the absence of the field. If $N(E)$ is the density of occupied states, then as we have shown in section 4.4.5,

$$N(E) = 2D(E)f(E)$$

where $f(E)$ is the probability of a given state of energy E being occupied, and $D(E)$ is the density of the states (i.e. the number of states with the same energy E). Clearly $0 \leqslant f(E) \leqslant 1$ where f is the Fermi function.

The Pauli paramagnetic susceptibility $\chi_P = M/H$ is therefore given by

$$\chi_P = 2m \left(\frac{dN_0}{dE}(E_F) \right) \frac{\Delta E}{H}$$

and since $\Delta E/H = \mu_0 m$,

$$\chi_P = 2D(E_F)\mu_0 m^2$$

where $D(E_F)$ is the density of states at the Fermi level and m is the magnetic moment of an electron.

It is clear that the Pauli paramagnetic susceptibility is dependent entirely on the small fraction of electrons which reside close to the Fermi level. Typically it is found that $\chi_P \approx 10^{-5}$ (dimensionless).

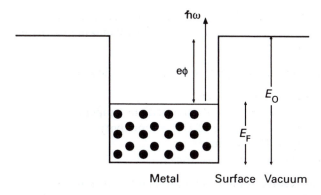

Fig. 4.11 Square-well potential model of a metal. The work function ϕ is equal to the difference in energy between the Fermi level and the energy of an electron outside the material.

4.5.3 Thermionic emission

What happens when a metal is heated to very high temperatures?

Finally we will mention in passing the phenomenon of thermionic emission, which again confirms our basic understanding of the behaviour of electrons in metals. Electrons do not escape from a metal very easily. So their energies outside the metal must be considerably higher than their energies inside. We can represent the metal as a finite potential well for the electrons (Fig. 4.11).

We define the work function ϕ as the minimum extra energy, measured above the Fermi level E_F, which an electron must obtain in order for it to escape from the solid. If E_0 is the energy of the electron outside

$$\phi = E_0 - E_F.$$

Once the temperature of the metal is raised above zero the sharp distinction between occupied and unoccupied states at the Fermi level becomes blurred as more electrons are thermally excited to higher energy states. For most metals a typical value of the work function is $\phi = 4\,\text{eV}$, as shown in Table 3.2 but for some oxide-coated filaments it is as low as $2\,\text{eV}$. It is clear that we must have $k_B T > \phi = E_0 - E_F$ to observe thermionic emission.

4.6 CONCLUSIONS

In what ways does the model succeed or fail to give good predictions of the properties of materials?

The Sommerfeld quantum mechanical model had some notable successes, particulary the explanation of the electronic contribution to the heat capacity, but also ultimately it had some failures. The model failed to explain the distinction between metals and insulators or semiconductors. It also failed to explain the number of conduction electrons by making no distinction between the bound and free electrons in the solid.

In addition it did not adequately explain the following:

– the value of the Hall coefficient;
– the Wiedemann–Franz law at intermediate temperatures;
– the temperature dependence of DC conductivity;
– the anisotropy of DC conductivity;
– the number of conduction electrons.

In order to resolve some of these difficulties it is necessary to consider the electrons in the presence of a periodic potential instead of a flat square-well potential. Certain electrons are then trapped within the periodic potential wells around the ionic cores while other higher energy electrons ride above the local potential wells and are constrained only by the boundaries of the material.

The periodic potential within the material is much more realistic of course, because the electrostatic interaction between the atomic cores at the lattice sites and the electrons, which the model neglects, will lead to periodic spatial fluctuations in the potential which are not allowed in the flat internal potential of the Sommerfeld model. These have been already described classically in Chapter 2.

Metals are those materials in which there are a number of high-energy electrons riding above the periodic potential in quasi-free states. Insulators are materials in which the highest energy states are constrained within the localized ionic potentials. This means that the insulators do not have 'free' electrons inside the material and therefore the highest energy electrons in insulators cannot contribute to the conductivity in their ground state.

REFERENCES

1. Schrödinger, E. (1926) *Ann. der Physik*, **79**, 361.
2. Sommerfeld, A. (1928) *Z. fur Physik*, **47**, 1.
3. Sommerfeld, A. and Bethe H. (1933) *Handbuch der Physik*, **24**, II.
4. Pauli, W. (1925) *Z. Physik*, **31**, 765.
5. Maxwell, J. C. (1866) *Phil. Mag.*, **32**, 390.
6. Boltzmann, L. (1868) *Sitz. Math–Naturwiss Cl. Akad. Wiss (Wien)*, **58**, 517.
7. Fermi, E. (1926) *Z. fur Physik*, **36**, 902.
8. Dirac, P. A. M (1926) *Proc. Roy. Soc. Lond.*, **112**, 661.
9. Goudsmit, S. and Uhlenbeck, G. E. (1926) *Nature*, **117**, 264.
10. Gerlach, W. and Stern, O. (1924) *Ann. Physik*, **74**, 673.
11. Pauli, W. (1926) *Z. Physik* **41**, 81.

FURTHER READING

Ashcroft, N. W. and Mermin, N. D. (1976) *Solid State Physics*. W. B. Saunders, Philadelphia.
Bube, R. H. (1971) *Electrons in Solids: An introductory survey*. Academic Press, San Diego.

MacDonald, D. K. C. (1956) Electrical conductivity of metals and alloys at low temperatures, in *Handbook of Physics*, Ed. S. Flügge, Springer Verlag, Berlin, p.137.

EXERCISES

Exercise 4.1 Fermi energy for a free-electron metal. Discuss the principal differences between the classical free-electron model and the quantum free-electron model. If the electrons described by the wave equation are completely free, determine the number density of electron states per unit volume with wave vector between k and $k + dk$, assuming that there are $1/4\pi^3$ states per unit volume of k-space. From this expression determine the number of states per unit volume of material with energy between E and $E + dE$ and calculate the Fermi energy for a free-electron distribution with n electrons per unit volume.

Exercise 4.2 Solution of wave equation in a finite square well. Derive the solution for the wave equation in a one-dimensional finite square-well potential of height V_0. What happens to the solutions if the electrons encounter a periodic potential within the square well? Describe the different types of energy states permissible under these conditions. Why are only certain energy levels allowed for 'quasi-free' electrons in the Sommerfeld model?

Exercise 4.3 Electronic specific heat of copper at 300 K. Using the quantum free-electron model calculate the electronic specific heat of copper at 300 K. At what temperatures are the electronic and lattice specific heats of copper equal to one another? (assume $\theta_D = 348$ K)

5
Bound Electrons and the Periodic Potential

We now look at the situation in which the wave equation is solved in the presence of a periodic potential. The result is a qualitative change in the form of solutions. In this chapter we focus exclusively on the electron properties as defined by the allowable solutions of the wave equation under these conditions. The lattice is present only to the extent that it provides a background for finding the allowed energy levels for the electrons. An important development of the model that arises as a result of the periodic potential is that the electrons inside the material are separated into two types: low energy 'bound' electrons which are spatially constrained to occupy the localized energy wells, and 'free' electrons which have higher energy and can migrate throughout the material. The result is allowed energy 'bands' separated by disallowed energy 'gaps'. The higher energy 'free' electrons are in general terms the same as the electrons described in the Sommerfeld model. The 'bound' electrons here represent a new addition to the quantum mechanical description of the electrons in the material.

5.1 MODELS FOR DESCRIBING ELECTRONS IN MATERIALS

What conceptual models do we have for describing the behaviour of electrons in materials?

So far we have dealt with the description of the electrons in solids in extremely simple terms. We have looked at:

(a) classical particles in a box
(b) waves in free space
(c) waves in an infinite potential
(d) waves in a finite potential.

Although each of these models is useful in as much as it gives us a general understanding of the possible behaviour of electrons in a solid, it must be admitted that all of the above are approximations which are likely to be far removed from the situation inside a real material. A real electron in a real solid will see the periodic potential formed by the atoms on the lattice sites. Ultimately this can hardly be described by the flat potential within the solid which is used in the square-well potential calculations.

5.1.1 Electrons in a periodic potential

What is the next extension of the model to bring it closer to the description of a real material?

The next level of complexity, which brings us surprisingly close to the real conditions in a solid, is to invoke a periodic potential within the solid. This was first attempted by Bethe [1] and Brillouin [2] who expanded the wave functions for an electron in terms of a series of plane waves subjected to the periodic potential. For calculation purposes, these can be square wells as shown, although in practice they will be more like parabolic potentials.

We have already established the methods for finding the allowed wave functions in this situation:

1. identify the boundary conditions (remember that both ψ and $d\psi/dx$ must be continuous across the boundaries)
2. solve the wave equation (in accordance with the boundary conditions).

In one dimension the simplest representation of a periodic potential has square potential wells, as shown in Fig. 5.1.

In this case there are three different energy ranges of interest

(a) $E < V_1$
(b) $V_1 < E < V_2$
(c) $E > V_2$

The result of subjecting the electrons to a periodic potential is that the energy levels become separated into allowed energy bands where electron states exist, and forbidden band gaps where no electron states exist.

5.1.2 High-energy free electrons

What are the solutions of the equation of state for very high energy electrons in the volume occupied by the material?

Clearly for $E > V_2$ we have the case of free electrons so that the wave equation can be solved without the electrons being constrained within the potential well. This means that all values of k are allowed and the energy spectrum is continuous. This is similar to the solution for the finite square-well potential in section 4.3.2 when $E > V_0$. In this case the electrons have escaped from the material.

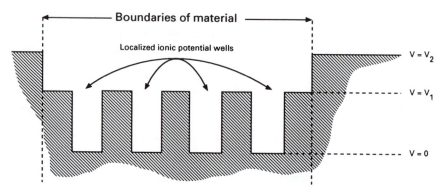

Fig. 5.1 One-dimensional representation of a periodic potential using simplified square wells.

5.1.3   Low-energy bound electrons

What form do the solutions of the equation of state take for low-energy electrons?

For $E < V_1$ we have a number of finite square-well potentials of width a. Electrons with these energies occupy bound states within these local periodic potentials. This is similar to the situation described above for the single finite square-well potential with $E < V_0$. The electron states in this case correspond to those of electrons localized at the ionic cores on the lattice sites in a solid and consequently these electrons cannot take part in conduction. These are discrete energy levels.

5.1.4 Intermediate energy conduction electrons

What form do the solutions of the equation of state take for electrons of intermediate energy?

For $V_1 < E < V_2$ we have a new situation (Fig. 5.2). Here the boundary conditions are provided by the boundaries of the material. Since this length is much greater than the width of the local potential wells, this ensures that once the energy is above V_1 the difference in energy between successively higher allowed energy states is less than for the energy levels below V_1. The solutions here are similar to those of a finite square-well potential of depth $V_2 - V_1$ and for electrons with energy $E - V_1$. These are the conduction electrons. However, the periodic potential does perturb the solution of the equations even at these higher energies. The degree of perturbation depends on the depth, width, and number of periodic potentials within the box.

   The electrons in this energy range we should correctly call 'quasi-free'. That is they are constrained only by the physical limitations of the material. They are equivalent to the electrons described by the Sommerfeld model.

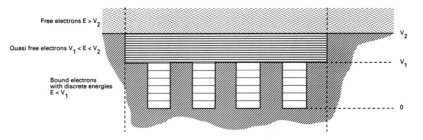

Fig. 5.2 Allowed electron energy levels in a one-dimensional 'square-well' lattice obtained by solving the Schrödinger equation.

5.1.5 Comparison with Sommerfeld free-electron model

How do the solutions differ from the Sommerfeld model predictions?

There are now three groups of electron states with qualitatively different properties. First there are the totally free electrons whose energies are large enough for them to have completely escaped from the metal. These will be of little intrinsic interest however. Next the quasi-free conduction electrons which are constrained only by the boundaries of the solid and form an almost continuous spectrum of k values or allowed energy states. Finally, there are the bound electrons whose energies are low enough for them to be trapped in the potential wells close to the atomic cores on the lattice sites. For these electrons the allowed energy levels are more widely separated.

It is the second group, the quasi-free electrons, which contributes to the electrical and thermal conductivity of metals. These electrons can move throughout the entire solid. We can see now, from this model, why a free-electron theory such as the Sommerfeld model seems to account so well for the bulk electrical and thermal properties of a metal. It is because the solutions of the wave function in the Sommerfeld model are quite close to the wave functions of the higher energy electrons in a metal, which are the main contributors to the electrical and thermal properties of metals. We shall find that the differences between a metal and an insulator is that the metal has some of these quasi-free electrons whereas the insulator does not. However, some refinements of our model will be needed later to allow semiconductors to have some of these higher energy electrons in filled 'energy bands' which cannot contribute to electrical conduction.

5.2 SOLUTION OF THE WAVE EQUATION IN A ONE-DIMENSIONAL PERIODIC SQUARE-WELL POTENTIAL

What changes occur in the solution of the wave equation in the presence of a periodic square-well potential?

We now introduce the periodicity of the crystal lattice in one dimension and solve the wave equation under these conditions as in the paper by Kramers

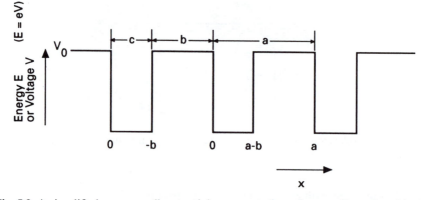

Fig. 5.3 A simplified square-well potential representation of a one-dimensional lattice.

[3]. Consider square-well potentials of height V_0, width c, periodicity a and distance between wells b. Then $b = a - c$ and the potential has the form shown in Fig. 5.3.

As we have shown before, the allowed energies can be found by solving the Schrödinger equation for the regions where $V = 0$ and $V = V_0$ separately. Spatially we separate the regions $V = V_0$ for $-b \leqslant x \leqslant 0$, and $V = 0$ for $0 \leqslant x \leqslant a - b$, and furthermore, $V(x + a) = V(x)$. The Schrödinger equation for the one-dimensional, time-independent case is,

$$-\frac{\hbar^2}{2m}\frac{d^2}{dx^2}\psi(x) + V(x)\psi(x) = E\psi(x)$$

which we now solve in the two regions of interest.

(a) Where $V = 0$,

$$\psi_0(x) = A\exp(i\beta x) + B\exp(-i\beta x)$$

and since $V = 0$, then the wave vector β is given by

$$\beta = \frac{(2mE)^{1/2}}{\hbar}.$$

(b) Where $V = V_0$

$$\psi_v(x) = C\exp(\alpha x) + D\exp(-\alpha x)$$

and since $V = V_0$, and assuming that $E < V_0$ we obtain the relation

$$\alpha = \frac{[2m(V_0 - E)]^{1/2}}{\hbar}.$$

So, if $E > V_0$ this leads to oscillatory (wave-like) solutions, but if $E < V_0$ the solutions are simply decaying exponentials.

Applying boundary conditions, it must be remembered that two continuity conditions must be satisfied:

(a) $\psi(x)$ must be continuous
(b) $d\psi(x)/dx$ must be continuous.

So for example, at $x = 0$ at the boundary of one potential well, our boundary conditions require that the wave function ψ_0 in the $V = 0$ region matches the wave function ψ_v in the $V = V_0$ region,

$$\psi_0(0) = \psi_v(0)$$

and also that the derivatives must match at the boundary

$$\frac{d}{dx}\psi_0(0) = \frac{d}{dx}\psi_v(0).$$

There is also a periodicity condition on the wave function. In order that the wave function should have the same periodicity as the lattice we also require that the wave function at $x + a$ is the same as at x. This periodicity condition can be expressed in terms of a wave function with vector k, as was shown by Bloch [4]

$$\psi(x + a) = \psi(x)\exp(ika).$$

Therefore at the boundaries $x = -b$ and $x = a - b$ we have

$$\psi_v(-b) = \exp(-ika)\psi_0(a - b)$$

$$\frac{d\psi_v}{dx}(-b) = \exp(-ika)\frac{d\psi_0}{dx}(a - b).$$

The above four conditions lead to four simultaneous equations in the unknowns A, B, C and D.

$\psi(0)$: $\qquad A + B = C + D$

$\dfrac{d\psi}{dx}(0)$: $\qquad i\beta(A - B) = \alpha(C - D)$

$\psi(-b)$: $\qquad \begin{aligned} C\exp(-\alpha b) \\ + D\exp(\alpha b) \end{aligned} = \exp(-ika)\left[\begin{aligned} A\exp(i\beta(a - b)) \\ + B\exp(-i\beta(a - b)) \end{aligned}\right]$

$\dfrac{d\psi}{dx}(-b)$: $\qquad \begin{aligned} \alpha C\exp(-\alpha b) \\ - \alpha D\exp(\alpha b) \end{aligned} = \exp(-ika)i\beta\left[\begin{aligned} A\exp(i\beta(a - b)) \\ - B\exp(-i\beta(a - b)) \end{aligned}\right]$

These four simultaneous equations can be solved if we require the determinant of the coefficients A, B, C and D to vanish. In that case the following energy restriction is obtained

$$\cos(ka) = \left(\frac{\alpha^2 - \beta^2}{2\alpha\beta}\right)\sinh(\alpha b)\sin(\beta(a - b)) + \cosh(\alpha b)\cos(\beta(a - b))$$

The limitation imposed by this condition is that the expression on the right-hand side of the equation must lie between ± 1 for allowed solutions. Mathematically it is possible for the expression on the right-hand side to lie outside the range ± 1, but then this does not correspond to a physically allowed solution.

5.2.1   Kronig–Penney approximation

What simple approximations can be made to the periodic potential in order to demonstrate the form of the solutions?

A simplification of the above constraint can be obtained by a mathematical model known as the Kronig–Penney approximation [5]. In this the width of the potential barriers, b, is allowed to decrease to zero, while the height V_0 of the barriers is allowed to increase to infinity under the condition that the product bV_0 remains constant. In this case then we have the following limits:

$$\lim_{b \to 0} \sin \beta(a - b) = \sin \beta a$$

$$\lim_{b \to 0} \cos \beta(a - b) = \cos \beta a$$

$$\lim_{b \to 0} \sinh \alpha b = \alpha b$$

$$\lim_{b \to 0} \cosh \alpha b = 1.$$

The following relations for the wave vectors still hold,

$$\beta = \frac{(2mE)^{1/2}}{\hbar}$$

$$\alpha = \frac{(2m(V_0 - E))^{1/2}}{\hbar}$$

and again for electrons in the regions with potential $V = V_0$ and with energies greater than V_0 solutions are wave-like, while for those in the same regions with energies less than V_0, α is real and the solutions are exponentials.

Now the previous constraint becomes as follows

$$\cos(ka) = \frac{P}{b\alpha} \sin(\beta a) + \cos(\beta a)$$

where

$$P = \frac{mab V_0}{\hbar^2}.$$

Allowed energy bands according to this model are shown in Fig. 5.4.

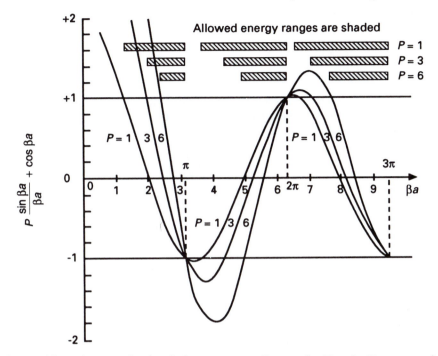

Fig. 5.4 Allowed energy levels of electrons according to the Kronig–Penney model. Reproduced with permission of R. H. Bube, *Electrons in Solids: An introductory survey*, published by Academic Press, 1971.

The only energies allowed are those for which

$$\left| \frac{P}{b\alpha} \sin(\beta a) + \cos(\beta a) \right| \leqslant 1.$$

As βa increases with the lattice parameter a, the allowed energy bands become wider because the solutions become more like those of free electrons. However as the potential V_0 increases, so $P(= abmV_0/\hbar^2)$ increases, the electrons become less free and the allowed energy bands become narrower. As V_0 increases further the solutions ultimately will resemble those of an electron trapped in an infinite square-well potential.

5.2.2 The 'nearly free' approximation

What values of energy are allowed for a wave in a one-dimensional discrete periodic lattice?

Another way of looking at the behaviour of electrons in a periodic potential is to consider the allowed solutions of the wave equation in a periodic potential as a perturbation from free-electron behaviour. Using a one-dimensional lattice of length L and with n atoms it can be shown that the allowed values of k are governed by

$$k_n^2 = \frac{\pi^2 n^2}{L^2}$$

and since for free electrons $E = (\hbar^2 k^2/2m)$, it is immediately apparent that only certain restricted values of energy are allowed by the lattice: those corresponding to integer values of n

$$E_n = \frac{\hbar^2}{2m}\frac{n^2\pi^2}{L^2} = \frac{n^2 h^2}{8mL^2}.$$

Extending this to three dimensions gives

$$k_n^2 = k_x^2 + k_y^2 + k_z^2$$
$$= \frac{\pi^2}{L^2}(n_x^2 + n_y^2 + n_z^2)$$

and

$$E_n = \frac{\hbar^2}{2m}(k_x^2 + k_y^2 + k_z^2).$$

The idea of solving the wave equation in an 'empty lattice' has been discussed by Shockley [6] and provides a limiting case for very weak potentials.

5.2.3 Density of states

How does the number of available energy states for electrons vary with energy of the electrons in the free-electron approximation?

The number of different k states that is possible up to a given value k_n can be calculated easily in the free-electron limit by noting that the values n_x, n_y and n_z must all be positive. Therefore the number of available states is simply the number of unit cubes in the positive quadrant of n-space of radius n. In other words it is one eighth of the volume of a sphere in n-space

$$N_0(E_n) = \frac{1}{8}\cdot\frac{4}{3}\pi n^3$$

and since $n = Lk_n/\pi$

$$N_0(E_n) = \frac{1}{6\pi^2}L^3 k_n^3.$$

Here we must remember that the electrons have spin, and therefore each energy level can be occupied by two electrons. Therefore the number of electrons from $k = 0$ to $k = k_n$ will be twice $N_0(\dot{E}_n)$

$$2N_0(E_n) = \frac{1}{3\pi^2} L^3 k_n^3.$$

If we wish to express this number of electron states in terms of energy E_n instead of k_n, we need to know the relationship between E_n and k_n. In general this is not known, but if we make the free-electron approximation, $E(k) = \hbar^2 k^2/2m$, then the number of available states between $E = 0$ and $E = E_n$ is

$$2N_0(E_n) = \frac{1}{3\pi^2} L^3 \left(\frac{2mE_n}{\hbar^2} \right)^{3/2}.$$

This function can then be differentiated with respect to E to obtain the density of states per unit energy interval

$$D(E) = \frac{\mathrm{d}}{\mathrm{d}E} N_0(E)$$

$$= \frac{L^3}{4\pi^2} \left(\frac{2m}{\hbar^2} \right)^{3/2} E^{1/2}$$

$$= \frac{V}{4\pi^2} \left(\frac{2m}{\hbar^2} \right)^{3/2} E^{1/2}$$

where V is the volume of the material. This is the same result that was obtained in section 4.4.7 beginning with a different condition, the square-well potential.

5.3   THE ORIGIN OF ENERGY BANDS IN SOLIDS: THE TIGHT-BINDING APPROXIMATION

How can we view the electron energy levels in a material as if they had evolved from the energy levels in single atoms?

We have already seen in the previous discussion that there is a qualitative difference in the properties of the high-energy, quasi-free electrons which are responsible for electrical and thermal conduction, and the low-energy bound electrons which are trapped in the atomic potential wells located on the lattice sites.

Now we shall take another approach originated by Bloch [7] which also contributes to our understanding of the properties of electrons in solids. It is well known that the energy levels of electrons within a single isolated atom are highly discrete. This was shown for example in the Bohr model of the atom, in which these energies were given by

$$E_n = - \left(\frac{me^4}{8h^2 \varepsilon_0^2} \right) \frac{1}{n^2}$$

where n is an integer, m is the mass of the electron, e is the charge on the electron, h is Planck's constant and ε_0 is the permittivity of free space.

If we consider what happens to these sharply defined electronic energy levels as the atoms of a solid are brought together we find that they become broadened. This occurs because the Pauli exclusion principle does not allow two electrons to have completely identical states. Therefore, the energies which were identical in the isolated atoms must shift relative to one another, and the closer the atoms

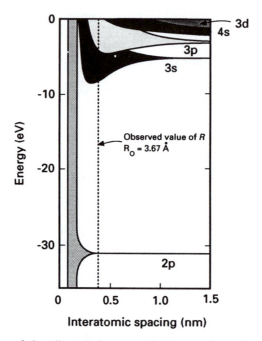

Fig. 5.5 Broadening of the allowed electron energy states into electron 'bands' as the interatomic spacing a is reduced. Reproduced with permission of R. H. Bube, *Electrons in Solids: An introductory survey*, published by Academic Press, 1971.

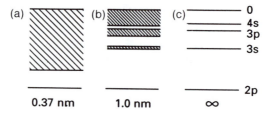

Fig. 5.6 Energy 'spectrum' of allowed states for: (a) free electrons outside a solid; (b) band electrons within a solid; and (c) bound electrons in isolated atoms. Reproduced with permission of R. H. Bube, *Electrons in Solids: An introductory survey*, published by Academic Press, 1971.

are together the more marked is the shift in available energy states. Furthermore, the number of different energy states is dependent on the number of atoms.

This means that a discrete energy level in a particular type of atom broadens into an allowed band of energies when a large number of identical such atoms are brought together in a solid (Fig. 5.5). The higher energy states broaden first as the atoms are brought closer together. The broadening of the lower energy states, which are closer to the atomic cores, is less marked.

If we look more closely at the three classes of electrons, the two extreme cases of bound electrons and free electrons and the intermediate case of electrons in a solid, we find the types of energy spectrum shown in Fig. 5.6.

5.3.1 The tight-binding approximation versus the free-electron approximation

How do the extreme viewpoints of single atom versus collective electrons compare?

The approach which we have just taken, considering the energy levels in the isolated atom and then bringing the atoms together and modifying the energy levels is known as the 'tight-binding approximation' because we start from tightly bound electrons in the atoms. The previous approach, beginning with the free-electron model and progressing through to a periodic potential model is known as the free-electron approximation. Both should lead to the same result if their conditions are relaxed sufficiently.

Both are valid methods of looking at the electronic properties of materials but approach the problem from different directions. In fact an intermediate model by Herring [8] has also been used to calculate the energy states of electrons in solids by combining the best aspects of both models. This is known as the orthogonalized plane-wave method. In metals, of course as we have already noted, the higher energy electrons behave as if almost free. Therefore, the free-electron approximation is more relevant in these cases.

In semiconductors and insulators, the electrons do not become free in the same sense because the highest occupied energy states in the ground state are localized at the atomic lattice sites, and so in this case the tight-binding approximation is more relevant. However, in either model we are led to the conclusion that when large numbers of atoms are brought together, energy bands arise and these bands are separated by forbidden energy regions or band gaps.

5.3.2 Transition from insulator to metal under pressure

What happens to the electronic properties of an insulator if it is subjected to extremely high pressure?

Since in their ground states electrons will always occupy the lowest available energy state, it is possible to convert an insulator to a metal under very intense

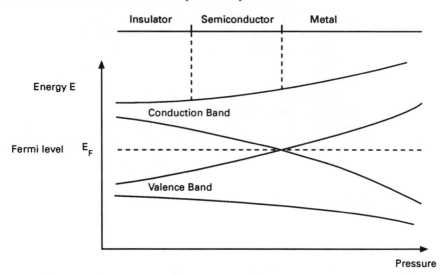

Fig. 5.7 Effects of pressure on the allowed electron energy levels of atoms. At low pressure the atoms are isolated. At an intermediate pressure they form an insulator and at very high pressures they form a metal. Compare with Figs 5.5 and 5.6.

pressure as a result of the broadening of the energy bands which occurs when the atomic cores are moved closer together as shown in Fig. 5.7.

If we make the assumption that the Fermi level does not change, then the material will undergo a transition from insulator to metal at the point where the two previously separated valence and conduction bands begin to overlap.

We should note that the pressures needed to cause this type of transition are very high. For example germanium, which is a semiconductor under normal conditions, becomes a metal under a pressure of 120 kbar (12 GPa). Recent results also suggest that solid hydrogen becomes metallic under a pressure of 2.5 Mbar (250 GPa).

5.4   ENERGY BANDS IN A SOLID

How are the properties of the electrons in the different energy bands qualitatively different from each other?

We have seen how the discrete energy levels in an atom are broadened into allowed energy bands when the atoms are brought together in a solid. The upper energy bands, which correspond to electrons with energies above the periodic potential wells of the atoms in the solid, contain the so called 'free electrons'. Below this there may be other free-electron type bands. Finally, below the free-electron bands are the bound states. The free-electron bands are constrained only by the boundaries of the solid. The bound electrons are constrained by the local potential wells around the atomic cores. The local ionic potentials cause a perturbation of the wave functions of the free-electron bands so that they are not identical to the solutions obtained in the Sommerfeld model.

5.4.1 Width of energy band gaps

How are the widths of the energy gaps related to the periodic potential?

Clearly since the energy gaps are caused by the presence of the periodic potential there must be some relationship between them. In fact the relationship is quite simple. The band gaps in the electron energy levels are equal to the Fourier coefficients of the crystal potential. These can be obtained by finding the Fourier series expansion of the periodic crystal potential and determining the coefficients of the series. This result can be understood because the electrons behave as waves under the influence of the periodic potential. It provides a very useful way of determining the band gaps in a periodic potential.

5.4.2 Electron band structure in conventional space

How can we view the various electron bands in real space?

In real space the electrons are confined within the solid. Low-energy 'bound' electrons cannot participate in conduction unless they are thermally excited and escape from the atomic core. High-energy 'quasi-free' electrons migrate throughout the solid, being constrained by the physical boundary of the solid only (Fig. 5.8). The movement of these higher energy electrons is almost unaffected by the periodic potential of the atomic cores.

So far we have looked only at the distribution of electrons in real space. We will introduce later the concept of reciprocal space which can be used to describe the electronic states of a solid in a particularly economical and elegant manner.

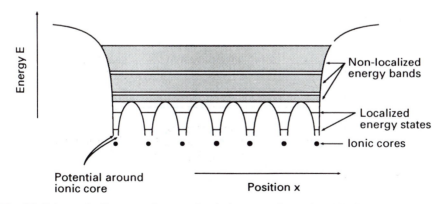

Fig. 5.8 Schematic diagram of energy levels in a one-dimensional lattice, shown in real space.

5.4.3 The Fermi energy

What is the highest energy level occupied by an electron when all electrons are in their lowest available state?

We have stated that at absolute zero, when the electrons all occupy the lowest available energy state, the energy of the highest occupied state is the Fermi level. This energy level separates the occupied from the unoccupied electron levels only when the electron configuration is in its ground state, that is, only at 0 K.

The location of the Fermi level in relation to the allowed energy states is crucial in determining the electrical properties of a solid. Metals always have a partially filled free-electron band, so that the Fermi level corresponds to a level in the middle of the band and this makes the metals electrical conductors. Semiconductors always have completely filled or completely empty electron bands. This means that the Fermi energy lies between the bands, and consequently they are poor electrical conductors at ambient temperatures.

5.4.4 Nomenclature of electron bands

How can the most important energy bands be described in a distinct and self-consistent manner?

We shall define the highest energy electron band containing electrons when the material is in its ground state as the valence band. We shall define the lowest energy band containing unoccupied electron states when the material is in its ground state as the conduction band, since it is through this band that electrical conduction can take place. In a semiconductor or insulator, the distinction is clear and we can represent the bands as illustrated in Fig. 5.9.

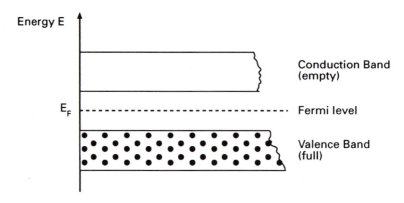

Fig. 5.9 Schematic diagram of electron energy bands in a semiconductor or an insulator.

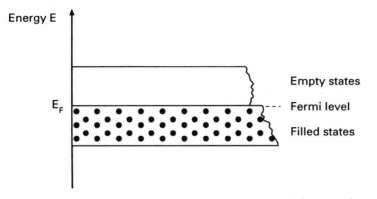

Fig. 5.10 Schematic diagram of electron energy bands in a metal.

In a metal which contains a partially filled band, this band satisfies both criteria and so is both a valence band and a conduction band (Fig. 5.10). This explains some of the confusing nomenclature of the bands in metals which occurs in the literature, in which the free electrons are sometimes described as 'conduction' electrons and sometimes as 'valence' electrons. In a metal they are both.

5.4.5 Effective mass of electrons in bands

How can the motions of electrons within an energy band be described in a simple way?

It is found experimentally that the mobility of electrons in the conduction band is affected by how full the band is. In the case of free electrons we have shown that,

$$E(k) = \frac{\hbar^2 k^2}{2m}$$

where m is the mass of the electrons, which is a constant in this case.

As we shall see shortly, this relationship between E and k breaks down in a solid. However, we can maintain the form of this relationship by using the relation,

$$E(k) = \frac{\hbar^2}{2m^*} k^2$$

where now m^* is an adjustable parameter which we call the effective mass. This means that any deviations from a parabolic relationship between E and k can be expressed as a change in the effective mass of the electrons at that point in k-space. Remember that of course this is merely a convenient artifice which allows us to describe the behaviour of the electrons in bands. The electron does not actually change its mass, it is simply an expression of the changed relationship between E and k.

The effective mass m^* can be smaller or larger than the free-electron mass m. The cause of this apparent change in the mass of the electrons is the interaction between the electrons and the lattice in the material. Collisions between drifting electrons and atomic sites will slow down the acceleration of an electron which will lead to an increase in its effective, or apparent, mass.

Another way to view the situation is in terms of the curvature of the energy bands in k-space. This can be interpreted in the following way: using the relationship between energy E and wave vector k given in section 4.2 and taking

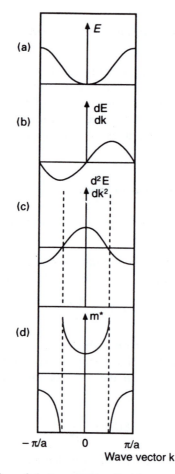

Fig. 5.11 Idealized variation of electron energy E with wave vector k. Diagram (c) shows the curvature of the electron energy band giving large effective mass at intermediate k, but small effective mass at low k as shown in (d). At k values close to $\pm \pi/a$ the effective mass of the electrons is negative.

the second derivative of the energy with respect to the wave vector gives $\hbar^2/m$. This relationship only holds exactly for a free-electron parabola. When the relationship between E and k is no longer parabolic the deviation can be expressed in terms of a change in the effective mass so that,

$$\frac{d^2E}{dk^2} = \frac{\hbar^2}{m^*}$$

and consequently the effective mass can be defined by,

$$m^* = \frac{\hbar^2}{(d^2E/dk^2)}$$

where d^2E/dk^2 is the curvature of the electron levels or electron band in k-space.

For electron bands with high curvature m^* is small, while for bands with small curvature, that is flat electron bands, m^* is large. It is also worth noting that since d^2E/dk^2 can be negative m^* can be negative. This simply means that when an electron goes from state k to state $k + \delta k$ the momentum transfer to the lattice is greater than the momentum transfer to the electron. The electron therefore appears to have a negative mass.

As a simple example, consider the energy states, as shown in Fig. 5.11. Small effective masses occur at low k in this case, large effective masses occur at intermediate k. Since the effective mass is determined by the curvature of the energy band in k-space, this means that narrow bands necessarily contain electrons with high effective mass. Conversely wide bands can contain electrons of low effective mass or high effective mass.

5.5 RECIPROCAL OR WAVE VECTOR k-SPACE

Is there an economical way of describing all of the allowed energy states in a solid?

Earlier in section 4.2 we introduced the idea of the wave vector k. This arose when we made a simple calculation for the solution of the wave equation for free electrons and then for electrons in a square-well potential. A plot of energy E against wave vector k was given first in Fig. 4.1. The dimensions of k are reciprocal length. It tells us the spatial periodicity of the wave function, or if you prefer, the number of cycles of the wave which occur in a given distance of 2π metres.

We found that for free electrons the energy depended on k^2, and all values of k were allowed. When the electrons are trapped, as in the square-well potential, only certain values of k are allowed in order that the wave functions can meet their boundary conditions.

In the last section we began to plot energy against k because this was useful in determining the effective mass of the electrons. We shall find that when it comes to describing electrons in solids plotting E against k is a very useful way of representing the electronic properties of the material. The plot of E against k is known as a reciprocal space plot because the dimensions of k are metre^{-1}.

When an electron is confined within a solid and experiences the periodicity of the lattice this periodicity affects the relationship between E and k. Another way of looking at this is that a wave function described by the wave vector k will have different energies depending on the presence and type of the crystal lattice it encounters. We have already noticed, for example, that the interactions between an electron and the lattice alter its effective mass and so distort the relationship between E and k.

5.5.1   Brillouin zones

How can periodicity of the lattice be introduced into reciprocal space?

We may think of the Brillouin zones [9] as a method for introducing the periodicity of the crystal lattice into our model of the electronic structure of materials. The Brillouin zone is a region in reciprocal space. Before saying more about exactly what this Brillouin zone is, let us consider the effects of a periodic potential on the energy levels of 'free electrons'.

We know $E = \hbar^2 k^2 / 2m$ for free electrons, but when we add the presence of a periodic potential all of this changes. We know from the previous chapter that if the energies of the electrons are below the level of the periodic potential barriers the electrons will penetrate spatially into these potential barriers, but their wave functions will be attenuated the further they penetrate. If the potential barriers are much higher than the electron energy, then the wave function will be reflected at these barriers and the electron will be contained entirely within the potential box formed by the barriers.

Now that we have introduced the idea of the electron wave function being reflected by the energy barriers of the periodic potential, let us find the condition for this. For Bragg reflection we need

$$2a \sin \theta = n\lambda$$

where a is the lattice parameter, θ is the angle of reflection, λ is the wavelength of the electrons and n is an integer. For simplicity let us again look at the one-dimensional lattice. In this case $\sin \theta = 1$ and the Bragg reflection condition is

$$2a = n\lambda$$

and $\lambda = 2\pi/k$ for an electron. So that

$$k = \frac{n\pi}{a}.$$

This then is the condition for reflection of the electron wave function in a one-dimensional lattice of parameter a. This can easily be generalized to three dimensions and the concept remains the same. We now have a division of reciprocal or k-space into a number of zones, known as Brillouin zones, at the boundaries of which reflection of the electron wave functions takes place.

The lattice can be divided into a number of Brillouin zones in reciprocal space:

First Brillouin zone $\quad\dfrac{-\pi}{a} \quad \text{to} \quad \dfrac{+\pi}{a}$

Second Brillouin zone $\quad\dfrac{-2\pi}{a} \quad \text{to} \quad \dfrac{-\pi}{a} \quad \text{and} \quad \dfrac{\pi}{a} \quad \text{to} \quad \dfrac{2\pi}{a}$

$\vdots \qquad\qquad\qquad\qquad \vdots$

nth Brillouin zone $\quad\dfrac{-n\pi}{a} \quad \text{to} \quad \dfrac{-(n-1)\pi}{a} \quad \text{and} \quad \dfrac{(n-1)\pi}{a} \quad \text{to} \quad \dfrac{n\pi}{a}$

Since free-electron wave functions are not reflected, and yet in a solid reflection occurs at the Brillouin zone boundaries, we may reasonably expect that the most severe deviation of the electron energies from free-electron-like behaviour will occur at the Brillouin zone boundaries.

In the plot of E versus k for a one-dimensional weak periodic potential shown in Fig. 5.12, the free-electron parabola is clearly apparent but with some distortion at the zone boundaries. At the k values corresponding to the zone

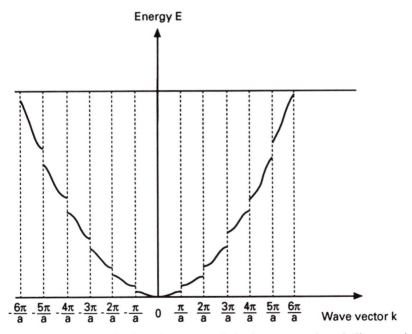

Fig. 5.12 Deformation of a free-electron parabola due to a weak periodic potential.

boundaries transmission of an electron through the solid is prevented. The incident and reflected electron wave functions form a standing wave. Certain energies are therefore forbidden since there are values of E for which there is no corresponding value of the wave vector k.

5.5.2   The reduced-zone scheme

Can the representation of the electrons in reciprocal space be made more compact by making use of the periodicity condition?

We can take our one-dimensional plot of energy versus wave vector and map all sections into the range $-\pi/a \leqslant k \leqslant \pi/a$. This is done by using the periodicity constraint so that,

$$k_1 = k_n + G$$

where k_n is the wave vector in the nth Brillouin zone, k_1 is the wave vector in the first Brillouin zone and G is a suitable translation vector. Now any point in k-space can be mapped by symmetry considerations to an equivalent point in the first Brillouin zone, but notice that many points from the extended-zone representation can be mapped to the same point in the reduced-zone representation.

This reduced-zone scheme representation allows the entire electron band structure to be displayed within the first Brillouin zone (Fig. 5.13). This is a very compact representation and has distinct advantages because the electronic states can be displayed in the most economical way in a single diagram of the first zone.

5.5.3   Band structures in three dimensions

How can information about the energy levels of a three-dimensional solid be represented in a compact manner on a two-dimensional diagram?

We have shown in previous sections the importance of plotting the electronic energy E against the wave vector k because this gives immediate information about the electronic properties of a material. We have also mentioned the reduced-zone representation which is useful because it presents this information in its most compact form. However, we have only done this so far for a one-dimensional lattice and a one-dimensional reciprocal lattice whose first Brillouin zone extends from $-\pi/a$ to $+\pi/a$.

In three-dimensional crystals with three-dimensional reciprocal lattices, the use of a compact representation is no longer merely a convenience, it is essential, otherwise the representation of the electronic states becomes too complex. How then can we display the band structure information from a three-dimensional crystal, which needs of course four dimensions (E, k_x, k_y and k_z) to describe it?

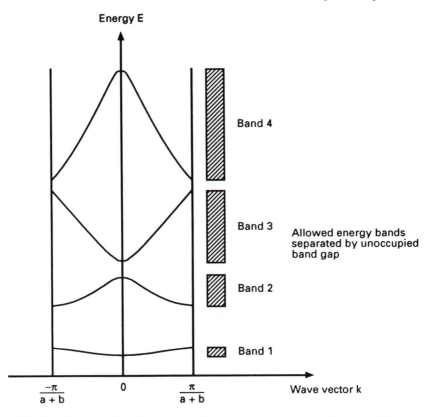

Fig. 5.13 The distorted free-electron parabola due to a weak periodic potential mapped back onto the first Brillouin zones.

The answer is to make representations of certain important symmetry directions in the three-dimensional Brillouin zone as one-dimensional E versus k plots. Only by doing this can we get all of our information onto a two-dimensional page. You can think of this process as cutting the Brillouin zone along certain symmetry directions. Therefore when looking at an E versus k diagram you are looking at several slices through different directions of k-space.

5.5.4 Brillouin zone of an fcc lattice

What does the 'unit cell' of an fcc lattice look like when transformed into reciprocal space?

The face-centred cubic lattice space group has been shown in Fig. 2.1. Now we will look at the Brillouin zone of such a lattice in k-space. This is shown in Fig. 5.14.

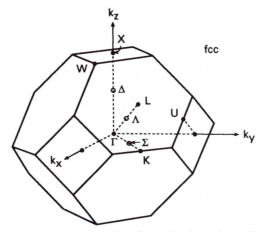

Fig. 5.14 Brillouin zone of an fcc lattice in reciprocal space.

Certain symmetry points of the Brillouin zone are marked. Specifically the Γ, X, W, K and L points and the directions Δ, Λ and Σ. Roman letters are used mostly for symmetry points and Greek letters for symmetry directions. The following is a summary of the standard symbols and their locations in k-space.

$$\begin{array}{ll} \Gamma & \langle 0,0,0 \rangle \\ X & \langle 1,0,0 \rangle \\ W & \langle 1,1/2,0 \rangle \\ K & \langle 3/4,3/4,0 \rangle \\ L & \langle 1/2,1/2,1/2 \rangle. \end{array}$$

5.5.5 Brillouin zone of a bcc lattice

What does the 'unit cell' of a bcc lattice look like when transformed into reciprocal space?

Similarly the Brillouin zone of a bcc lattice can be described in terms of its principal symmetry directions. The zone is shown in Fig. 5.15. The symmetry points are conventionally represented as Γ, H, P and N and the symmetry directions as $\Delta, \Lambda, D, \Sigma$ and G. The various symmetry points are

$$\begin{array}{ll} \Gamma & \langle 0,0,0 \rangle \\ H & \langle 1,0,0 \rangle \\ P & \langle 1,1,1 \rangle \\ N & \langle 1,1,0 \rangle. \end{array}$$

Notice that a bcc lattice in real space has a Brillouin zone in reciprocal space that is an fcc lattice, while an fcc lattice in real space has a Brillouin zone in reciprocal space that has bcc symmetry.

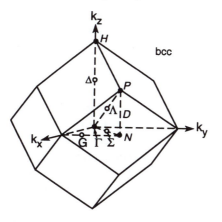

Fig. 5.15 Brillouin zone of a bcc lattice in reciprocal space.

5.6 EXAMPLES OF BAND STRUCTURE DIAGRAMS

What does the electron band structure diagram of the first Brillouin zone look like when represented in two dimensions?

Figures 5.16 and 5.17 show the electron band structures of two real materials, copper and aluminium. Aluminium is seen to contain electrons which in their

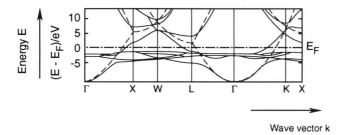

Fig. 5.16 Electron band structure diagram of copper.

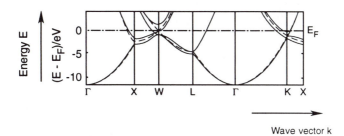

Fig. 5.17 Electron band structure diagram of aluminium. Reproduced with permission from B. Segal, *Phys. Rev.*, **125**, 1962, p. 109.

energy versus wave vector relationships are very close to free electrons, which would be represented by a free electron parabola in k-space.

5.7 CONCLUSIONS

In this chapter we have looked at the behaviour of electrons under the influence of a periodic potential due to the lattice ions. The calculations, although still oversimplified, do produce some interesting results such as the existence of localized low-energy states, and higher energy itinerant electron states with the two groups of states separated by a region of forbidden energy known as the band gap. Two different approaches to describing the electrons have been discussed, one beginning from modifying the energy states in isolated atoms, the other beginning with free electrons and modifying their behaviour with the effects of a periodic potential. It is possible to distinguish between conductors and insulators on the basis of this model, the conductors having electrons occupying the higher energy itinerant states, while the insulators have electrons confined to the localized energy states only.

We should now be in a position to understand the comment made in section 2.1, that the differences in the electronic properties of materials are due more to the ionic lattice than to the electrons themselves. Since all electrons are identical, the collective properties of the electrons are determined by the boundary conditions imposed on the electron wave function. The boundary conditions are periodic and are dictated by the ionic lattice. Without the lattice the electrons could have any energy value.

REFERENCES

1. Bethe, H. (1928) *Ann. Physik*, **87**, 55.
2. Brillouin, L. (1930) *J. Phys. Radium*, **1**, 377.
3. Kramers, H. A. (1935) *Physica*, **2**, 483.
4. Bloch, F. (1930) *Z. fur Physik*, **59**, 208.
5. Kronig, R. and Penney, W. G. (1931) *Proc. Roy. Soc.*, **A130**, 499.
6. Shockley, W. (1937) *Phys. Rev.*, **52**, 866.
7. Bloch, F. (1928) *Z. fur Physik*, **52**, 555.
8. Herring, C. (1940) *Phys. Rev.*, **57**, 1169.
9. Brillouin, L. (1931) *Quantenstatistik*, Springer, Berlin.

FURTHER READING

Bube, R. H. (1971) *Electrons in Solids: An Introductory Survey*, Academic Press, San Diego.
Coles, B. R. and Caplin, A. D. (1976) *The Electronic Structures of Solids*, Edward Arnold, London.
Mott, N. F. and Jones, H. (1936) *The Theory of the Properties of Metals and Alloys*, Oxford University Press.

EXERCISES

Exercise 5.1   Effective mass of electrons in bands. Explain the physical reasons why electrons in energy bands can behave as if they have different masses. Derive an expression for the effective mass and explain its significance in terms of the curvature of the electron bands. Do the electrons actually have a different mass?

Exercise 5.2   Origin of electron bands in materials. Explain how electron energy bands arise in materials, first beginning your discussion from the free electron approximation and secondly beginning your discussion from the tight-binding approximation.

The periodic potential in a one-dimensional lattice of spacing *a* can be approximated by a square wave which has the value $V = -2\,eV$ at each atom and which changes to zero at a distance of $0.1a$ on either side of each atom. Estimate the width of the first energy gap in the electron energy spectrum.

Exercise 5.3   Number of conduction electrons in a Fermi sphere of known radius. In a simple cubic, quasi-free electron metal the spherical Fermi surface just touches the first Brillouin zone. Calculate the number of conduction electrons per atom in this metal.

Part Two
Properties of Materials

6
Electronic Properties of Metals

In this chapter we bring together the basic concepts discussed in earlier chapters to provide a broad description of the electronic properties of metals. The main idea is that metals have some electrons which occupy the higher energy 'free' electron levels, and therefore can migrate throughout the material. These are the so-called conduction electrons which contribute to both electrical and thermal conduction. These electrons also enable us to explain the optical properties of metals, in particular the high reflectance of metals in the visual range of the spectrum. Since it is the electrons which are close to the upper surface of the electron 'sea' which are most important in defining the electronic properties, we look at this 'Fermi surface' in greater detail here and discuss some of the methods of representing this important electronic characteristic of a metal.

6.1 ELECTRICAL CONDUCTIVITY OF METALS

How do we account for the range of conductivities of materials?

We have mentioned the wide range of observed conductivities in materials ranging from $10^{-15}\,\Omega^{-1}\,m^{-1}$ in sulphur to $10^{8}\,\Omega^{-1}\,m^{-1}$ in copper. In order to discuss this range of properties systematically, we shall divide the materials into conductors and insulators, with the semiconductors such as germanium and silicon being classed with the insulators.

In this chapter we will look at the metals. Metals are typically good electrical and thermal conductors and good reflectors of light in the visible spectrum. These properties are all due to the free, or more precisely quasi-free, electrons in the material. Metals contain a partially filled electron band through which the electrons can move relatively freely in order to conduct heat and electrical current, and the electrons in this band can absorb photons of low energies simply by being excited to slightly higher, unoccupied electron levels within the same energy band, later returning to lower unoccupied levels with the emission of a photon.

Figure 6.1 shows a very simple schematic diagram of the highest occupied energy band in a metal. This schematic tells us nothing about the dependence

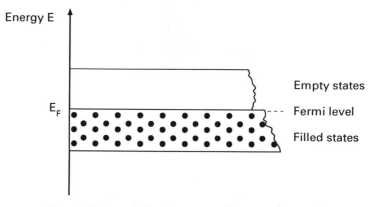

Fig. 6.1 Schematic band structure diagram of a metal.

of energy E on the wave vector k in this material. However we should note that the Fermi energy lies in the middle of an allowed band in a metal and this means that a metal has a Fermi surface, that is, an occupied connected set of highest energy levels in k-space.

6.2 REFLECTANCE AND ABSORPTION

How are the various optical properties of materials related to each other and to the electrical properties?

We have mentioned in section 1.4 that the optical properties of materials can be represented by two constants, either n and k or alternatively ε_1 and ε_2 as discussed in section 3.4.1.

The dielectric 'constant' of the material is $\varepsilon = \varepsilon_0(\varepsilon_1 + i\varepsilon_2)$, where ε_0 is the permittivity of free space. The two components of the complex relative dielectric constant $\varepsilon_r = \varepsilon_1 + i\varepsilon_2$ are related to n and k by

$$\varepsilon_1 = n^2 - k^2$$

$$\varepsilon_2 = 2nk.$$

Both ε_1 and ε_2 are dimensionless. The term ε_2 is known as the absorption. It is also related to the electrical conductivity $\sigma(\omega)$ by

$$\varepsilon_2(\omega) = \frac{\sigma(\omega)}{\omega\varepsilon_0}$$

at a given frequency of excitation ω. This expression was given by Drude in the classical theory of electrons in metals [1].

This means that a good electrical conductor, which has a high value of $\sigma(\omega)$ is a good absorber of light. A good electrical conductor is also known to be a good reflector of light (i.e. high R). This at first seems to be a contradiction. How can a good absorber also be a good reflector? We need to resolve this seeming contradiction immediately. The reflectance R is related to the components ε_1 and ε_2 of the dielectric constant (section 3.4.2) by the relation

$$R = \frac{\sqrt{\varepsilon_1^2 + \varepsilon_2^2} + 1 - \sqrt{2(\sqrt{(\varepsilon_1^2 + \varepsilon_2^2)} + \varepsilon_1)}}{\sqrt{\varepsilon_1^2 + \varepsilon_2^2} + 1 + \sqrt{2(\sqrt{(\varepsilon_1^2 + \varepsilon_2^2)} + \varepsilon_1)}}.$$

We know that high k leads to high ε_2 and high R. What do we mean by absorption in this context? We really mean that the light does not penetrate the solid very far. Clearly if the light is either reflected or transmitted, then with high absorption the light is not transmitted and so will eventually be reflected back. The mechanism for this process is absorption of light by free electrons, excitation of the electrons to higher energy states, and then de-excitation of the electrons with the emission of photons. In short, absorption followed by immediate reemission of the light gives reflection.

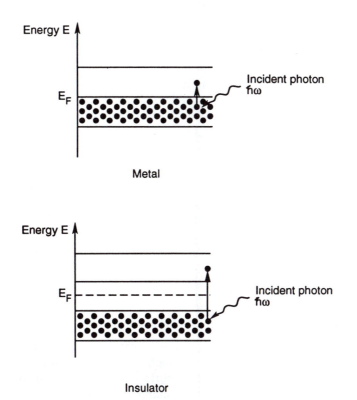

Fig. 6.2 Schematic metal and insulator band structure diagrams showing absorption of photons.

6.2.1   Optical properties and electron band structure

How can the optical properties be related to the electronic structure of the material?

Having stated this relation between absorption and reflectance we need to give a brief explanation in terms of the electron theory. In order to absorb incident light the electrons must have available states which they can move to at an energy ΔE above their present energy state, where

$$\Delta E = \hbar \omega$$

and ω is the frequency of the light and $\hbar$ is Planck's constant divided by 2π.

Immediately then, it is apparent than an insulator with a filled valence band, an empty conduction band and band gap $E_g > \hbar \omega$ cannot absorb the photon with energy $\hbar \omega$, because there are no available energy states into which the electrons could be excited which correspond to the correct gain in energy.

Metals, however, do have such energy states available at low energies because electrons can be stimulated to slightly higher available levels within the same band (intraband absorption). So the metals have a high absorption at low frequencies. Once the electrons in a metal have absorbed the light they can return to their available lower energy state with the emission of a photon of identical energy, or in some cases of lower energy. This leads to a high reflection coefficient R in metals.

Insulators on the other hand have low absorption at low frequencies, but as the energy of the incident radiation increases a frequency ω_0 is reached at which the energy of the incident photons equals the energy of the band gap

$$\hbar \omega_0 = E_g.$$

Beyond this frequency an insulator has a high absorption. Each semiconductor or insulator has its own characteristic energy gap E_g. This means that if we look at the reflection or absorption spectrum of a material we can soon tell whether it is a metal, semiconductor, or insulator, because high reflectance at low energies is a property of metals, whereas a low reflectance at low energies combined with high reflectance at high energies is a property of insulators.

6.3   THE FERMI SURFACE

How can we describe the Fermi level in three dimensions?

We have discussed the concept of a Fermi energy E_F and a Fermi level which is the highest occupied energy state in a metal in its ground state. Now we will generalize this idea further to the Fermi surface [2]. Excellent introductions to the Fermi surface have been given by Ziman [3] and by Mackintosh [4].

The Fermi surface is the plot of the Fermi level in three-dimensional k-space.

The volume contained within the Fermi surface represents all the occupied energy levels when the material is in its ground state.

In the simplest case the Fermi surface for free electrons is a sphere since the energy is given by an expression which depends on the sum of the squares of the k wave vectors in each direction.

$$E = \frac{\hbar^2}{2m}(k_x^2 + k_y^2 + k_z^2).$$

Since this expression for energy is not directionally dependent the Fermi surface for free electrons will be spherical in k-space.

In other cases, in particular in real metals, the Fermi surface is not exactly spherical in k-space, but for most metals which have upper electron states which are quasi-free, the Fermi surface is more or less spherical in shape when plotted in the extended-zone representation. There will of course be local perturbations due to the lattice, because of Bragg reflection at the Brillouin zone boundaries. This results in the Fermi surface having bulging contours close to the zone boundaries, as in the case of copper, silver and gold. The forms of these Fermi surfaces have been discussed in detail by Cracknell and Wong [5].

6.3.1 The Fermi surface in the reduced-zone scheme

Can the entire description of the allowed energy levels be represented in the first Brillouin zone alone?

As a consequence of the periodicity of the lattice, we have shown in Chapter 5 that the reciprocal lattice in k-space can be subdivided into Brillouin zones. In the one-dimensional case, the first Brillouin zone extends from $-(\pi/a)$ to $+(\pi/a)$ where a is the lattice parameter. A symmetry principle comes into play here, whereby the energy of a particular band $E(k)$ is a periodic function of the reciprocal lattice

$$E(k) = E\left(k \pm \frac{\pi}{a} \right).$$

This leads to the periodic zone scheme shown in Fig. 6.3 in which each energy state in a higher Brillouin zone can be mapped back to an equivalent point in the first zone.

The important result here is that the entire energy band structure of a material can be represented in the first Brillouin zone by transforming all the allowed energy states $E(k + n\pi/a)$ to the equivalent point in the first zone using the reduced-zone scheme

$$E\left(k + \frac{(2n-1)\pi}{a} \right) \quad \rightarrow \quad E\left(\frac{\pi}{a} - k \right)$$

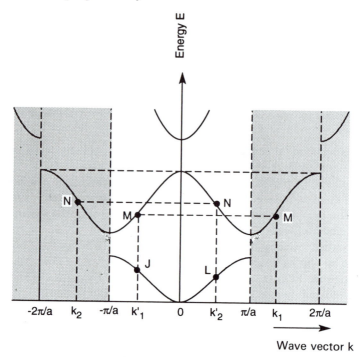

Fig. 6.3 Extended- and reduced-zone representations of electron energy levels.

$$E\left(k + \frac{2n\pi}{a}\right) \quad \rightarrow \quad E(k).$$

Therefore every point in every zone has an equivalent point in the first Brillouin zone.

6.3.2   Advantages and disadvantages of the reduced-zone scheme

Does the compact representation of the reduced-zone scheme lead to any disadvantages in visualizing the electronic structure of a material?

An obvious advantage of the reduced-zone representation is that the entire electron band structure of the material can be plotted within one Brillouin zone. This allows for compact representation of the electronic properties, particularly when we are dealing with three dimensions.

A disadvantage is that the electron energy levels, which may appear to be relatively simple in the extended-zone scheme can become very complicated to visualize in the reduced-zone scheme. A good example of this is the free-electron parabola which when mapped onto the first zone of an fcc lattice (in the absence

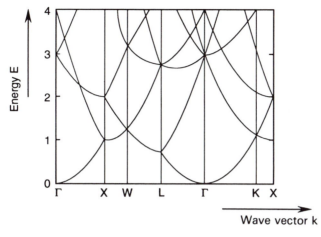

Fig. 6.4 Free-electron parabola represented in the reduced-zone scheme in the first zone of an fcc lattice, which shows how a very simple situation can appear to be superficially complicated under this transformation. Reproduced with permission from R. G. Chambers, *Electrons in Metals and Semiconductors*, published by Chapman & Hall, 1990.

of any periodic potential) is transformed as shown in Fig. 6.4. This corresponds to the 'empty lattice' approximation discussed by Shockley [6]. Despite the extremely simple nature of the relationship between E and k in the case of free electrons, the representation in Fig. 6.4 looks superficially quite complex. Weak Bragg reflection at the Brillouin zone boundaries can make this diagram appear even more complicated.

A Fermi surface which also may have a relatively simple distorted spherical shape in the extended-zone scheme can become an extremely complex shape in the reduced-zone scheme. An example of this is the Fermi surface of aluminium which is shown later (section 6.3.5).

6.3.3 Free-electron Fermi surface

What does the free-electron Fermi surface look like in the extended-zone scheme in two dimensions?

For simplicity we will begin by considering Fermi surfaces in a hypothetical two-dimensional solid. This is easier to represent and discuss. Once the general idea has been expounded we will go on to consider examples in three dimensions.

We have shown above that the Fermi surface of free electrons in k-space is spherical, and hence in two dimensions it is circular if we have no Bragg reflection (which is equivalent to an empty lattice). When the Fermi surface is entirely contained within the first Brillouin zone, the Fermi surface in the reduced-zone scheme is also circular. For simplicity we consider this situation in relation to a square lattice as shown in Fig. 6.5.

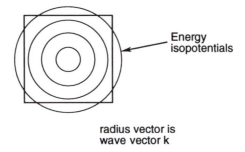

radius vector is
wave vector k

Fig. 6.5 Free-electron Fermi 'sphere' in two-dimensional *k*-space.

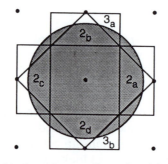

Fig. 6.6 Free-electron Fermi 'sphere' in two-dimensional *k*-space extending beyond the first zone.

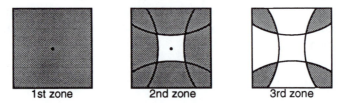

1st zone   2nd zone   3rd zone

Fig. 6.7 Free-electron Fermi 'sphere' mapped back into the first zone in two-dimensional *k*-space.

Now let us consider a metal with more electrons so that not all electron states can be contained in the first Brillouin zone. In this case in the extended-zone scheme the Fermi surface is again circular but the Fermi surface extends beyond the first zone (Fig. 6.6). Now in the reduced-zone scheme the situation appears more complicated.

When the free-electron Fermi sphere is folded back into the first zone in the reduced-zone scheme, the appearance of the surface is complicated by the transformation (Fig. 6.7).

The important point to note here is that for free electrons:

1. the Fermi surface in k-space in the extended scheme is perfectly spherical;
2. when projected back into the first zone using the reduced-zone scheme the spherical surface can look very different.

6.3.4   Fermi surface in a periodic potential

How does the presence of a periodic potential change the form of the Fermi surface?

Once we have introduced a weak periodic potential, the Fermi surface in k-space becomes distorted by Bragg reflection even in the extended-zone scheme. In our two-dimensional example the distortion from circular appears as shown in Fig. 6.8. The distortion from circular energy levels depends on the strength of the Bragg reflection at the zone boundaries, which alters the relationship between E and k.

It can be seen from the one-dimensional section of the E versus k diagram (Fig. 6.9) that an energy gap arises at the Brillouin zone boundary. This is a direct result of the periodic potential and the size of the gap depends on the strength of the periodic potential wells. The energy gap is caused by Bragg reflection of electrons with wave vectors close to the zone boundary. Clearly then, not all the electrons are reflected in this way.

If we only have a small number of valence electrons, for example, 1 per atom in the monovalent metals Cu, Au, Ag, then the Fermi surface is still relatively simple. Of course if the Fermi surface lies well inside the first zone (as in level 1 in Fig. 6.8) then the deviation from circular will be insignificant. However, as the band expands to fill the first zone and the Fermi surface gets closer to

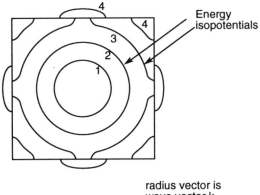

Fig. 6.8 Distorted two-dimensional Fermi 'sphere' in the first Brillouin zone resulting from the effect of a periodic potential.

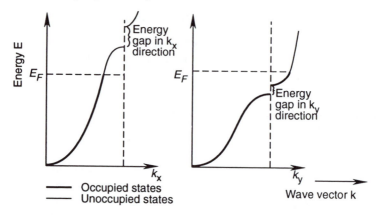

Fig. 6.9 One-dimensional electron energy levels in a rectangular first Brillouin zone showing band gaps at the zone boundary. If the reciprocal lattice is rectangular instead of square, the energy gaps occur at different energy levels in the different directions as shown here. This means that the Fermi surface can touch the zone boundary in one direction, the k_y direction, but not in the other direction, the k_x direction.

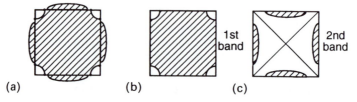

Fig. 6.10 Extended-zone representation of a two-dimensional distorted Fermi 'sphere': (a) extended-zone scheme; (b) representation in first zone; and (c) representation in second zone.

the zone boundary it will distort to meet the zone faces (e.g. level 3). Even larger numbers of electrons will push the Fermi level into the second zone (e.g. level 4).

Once we have parts of the Fermi surface in the second zone, we usually fold these back into the first zone to obtain a more compact reduced-zone representation. It is at this stage that the representation of an essentially rather simple Fermi surface can take on a very complicated appearance.

6.3.5   Three-dimensional Fermi surfaces of metals

How does the Fermi surface appear in three dimensions?

Having looked briefly at the problem of mapping the Fermi surface in two-dimensional k-space we now have a much better idea of what to expect in three dimensions when we look at the Fermi surfaces of real metals.

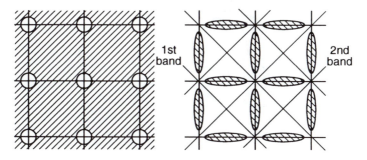

Fig. 6.11 Reduced-zone representation of a two-dimensional distorted Fermi 'sphere'.

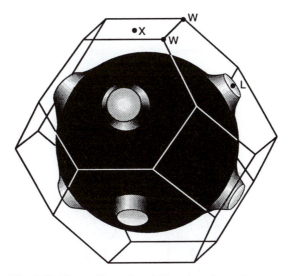

Fig. 6.12 Three-dimensional Fermi surface of copper.

The first metal for which the Fermi surface was completely mapped in k-space was copper. This work was performed by Pippard [7, 8]. The Fermi surface of copper is particularly easy to visualize because it all lies in the first Brillouin zone. From our discussion of the two-dimensional examples above, in which the Fermi surface that was contained entirely within the first zone was circular, we may expect the Fermi surface of copper to be approximately spherical. This is indeed the case, the surface does however have necks extending towards the Brillouin zone boundaries. This is caused by Bragg reflection where the Fermi surface comes close to the zone boundary. The shape of the Fermi surface resembles a 'diving-bell'. This is shown in Fig. 6.12. Both silver and gold have Fermi surfaces with similar shapes to that of copper.

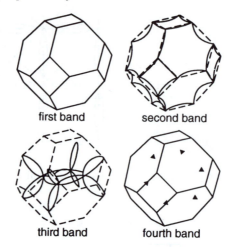

first band   second band

third band   fourth band

Fig. 6.13 Three-dimensional Fermi surface of aluminium.

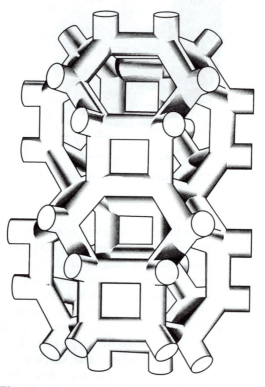

Fig. 6.14 Three-dimensional Fermi surface of lead.

In divalent calcium, which has of course two outer electrons per atom, the Fermi surface extends into the second Brillouin zone. Again, in the extended-zone scheme, the Fermi surface is approximately spherical. However, when folded back into the first zone in the reduced-zone scheme it looks quite different, resembling a 'coronet'. Divalent beryllium also has a coronet-shaped Fermi surface, however, since it solidifies with hexazonal symmetry the surface in reciprocal space has six-fold symmetry.

Aluminium, being trivalent, has three outer electrons. Its electron bands are very close to free-electron parabolae, and its Fermi surface if plotted in the extended-zone scheme is fairly simple in shape, being almost spherical and extending into the third Brillouin zone. Again once it is folded back onto the first zone it takes a very different form which has come to be known as the 'monster'. This is shown in Fig. 6.13.

Lead has four valence electrons and its Fermi surface extends into the fourth Brillouin zone. In this case the Fermi surface in the first zone is extremely complicated forming a 'pipeline maze', as shown in Fig. 6.14.

6.3.6 Methods of determining the Fermi surface

How can we examine the shape of the Fermi surface of a metal?

First we should say why measurements of the Fermi surface are important at all. Recall that the classical Drude metal of electrons in solids failed over the prediction of the specific heat capacity of metals. This was because even the free electrons in the conduction band of a metal are not able to absorb thermal energy unless they are within an energy $k_B T$ of the Fermi surface. This means that most of the electronic properties of a metal are determined by electrons lying at, or just below, the Fermi surface. Clearly the electrons close to the Fermi surface are most important in determining those properties of a metal which depend on the electrons. We conclude therefore, that by knowing the details of the Fermi surface we can make predictions about many of the properties of a metal.

There are a number of different measurements that can be made which give information about the Fermi surface. For a detailed description of these consult Ashcroft and Mermin [9] who have devoted an entire chapter to methods of measuring the Fermi surface. We list here only the most important techniques used:

- de Haas–van Alphen effect
- magnetoacoustic effect
- ultrasonic attenuation
- anomalous skin effect
- magnetoresistance
- cyclotron resonance
- positron annihilation.

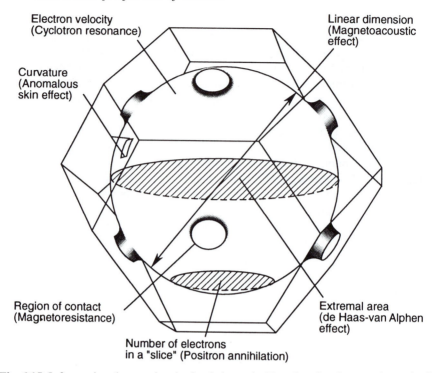

Electron velocity
(Cyclotron resonance)

Linear dimension
(Magnetoacoustic
effect)

Curvature
(Anomalous
skin effect)

Region of contact
(Magnetoresistance)

Extremal area
(de Haas-van Alphen
effect)

Number of electrons
in a "slice" (Positron annihilation)

Fig. 6.15 Information that can be obtained about the Fermi surface from various physical measurements as described by Mackintosh [4].

Each of these measurements gives different information about the Fermi surface. These are depicted in Fig. 6.15, based on a figure given by Mackintosh [4].

The de Haas–van Alphen effect [10] is the most important technique which is used to probe the Fermi surface. From these measurements the extremal area (i.e. largest cross-sectional area) of the Fermi 'sphere' can be found. The magnetoacoustic effect enables the linear dimension (i.e. largest diameter) of the Fermi 'sphere' to be calculated. In the absence of a magnetic field the conventional ultrasonic attenuation also gives information about the Fermi surface, however in this case the interpretation in terms of Fermi surface geometry is more complicated.

The anomalous skin effect, which can be used to measure the curvature of the Fermi surface in k-space is one of the oldest techniques used for Fermi surface measurements. It dates back to the work of Pippard [7, 8]. The penetration of the magnetic field into a solid at higher frequencies deviates from the classical skin effect equation and it can be shown that the field penetration becomes dependent entirely on certain features of the Fermi surface geometry (e.g. the curvature) at sufficiently high frequencies.

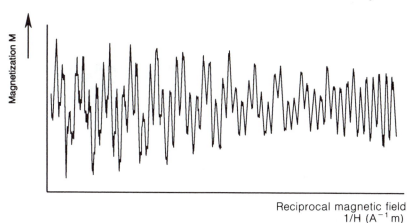

Reciprocal magnetic field
1/H (A^{-1}m)

Fig. 6.16 De Haas–van Alphen oscillations. Reproduced with permission from R. G. Chambers, *Electrons in Metals and Semiconductors*, published by Chapman & Hall, 1990.

Magnetoresistance measurements, that is the dependence of electrical resistance on magnetic field, can be used to find the region of contact of the Fermi surface with the Brillouin zone boundary since the magnitude of this contact affects the conductivity. Cyclotron resonance, the circular motion of a charged particle moving in a plane normal to a magnetic field, can also be used to investigate the Fermi surface. Specifically, it can be used to find the electron velocity on the Fermi surface.

Finally, positron annihilation can be used to find the number of electrons in a two-dimensional slice through the Fermi 'sphere'. When the material is bombarded with positrons the electrons annihilate the positrons, yielding two photons. The momentum of the emitted photons can be used to determine the momentum distribution of the electrons in the metal. This can then be used to indicate the number of electrons in a given slice through k-space.

6.3.7 The de Haas–van Alphen effect

How does the differential susceptibility of a metal depend on the applied field strength?

The de Haas–van Alphen effect is the most important technique used for obtaining information about the Fermi surfaces of metals. At low temperatures and under high magnetic fields (typically H > 5 kOe or 400 kA m^{-1}) it was found that the differential susceptibility dM/dH of metals was dependent on the field strength in an oscillatory manner.

When the differential susceptibility is plotted against $1/H$ this periodic dependence is shown most clearly, although often two or more periods are superimposed. Similar behaviour has been observed in the conductivity and the magnetostriction. The former is known as the Shubnikov–de Haas effect. Two methods are widely employed to measure these de Haas–van Alphen oscillations. One uses a torque magnetometer and simply measures the oscillations in angular position of a sample of the metal as the field strength H and magnetization M are increased.

The second method uses field pulses and measures the voltage induced in a flux coil wound on the sample. Since the voltage from the flux coil will be $V = N(\mathrm{d}B/\mathrm{d}t) \simeq \mu_0 N(\mathrm{d}M/\mathrm{d}t)$ and the rate of change of magnetic field $(\mathrm{d}H/\mathrm{d}t)$ is known, then it follows that,

$$V \simeq \mu_0 N \frac{\mathrm{d}M}{\mathrm{d}H} \frac{\mathrm{d}H}{\mathrm{d}t}$$

and so $(\mathrm{d}M/\mathrm{d}H)$ can be calculated. Onsager [11] has shown that the periodicity of the oscillations in $(1/H)$, measured in Oe^{-1}, is given by,

$$\Delta(1/H) = \frac{2\pi e}{\hbar} \frac{1}{A_{\mathrm{ext}}}$$

where A_{ext} is the extremal 'area' in reciprocal space of the Fermi surface in a plane normal to the magnetic field. So for a spherical Fermi surface $A_{\mathrm{ext}} = \pi k_{\mathrm{F}}^2$.

From measurements of the oscillations of $(\mathrm{d}M/\mathrm{d}H)$ against $(1/H)$ it can be seen that the extremal area is easily calculated from the periodicity, since π, e and $\hbar$ are all well-known constants

$$A_{\mathrm{ext}} = \frac{2\pi e}{\hbar\Delta(1/H)}.$$

This enables the extremal area of the Fermi surface to be determined in different directions.

REFERENCES

1. Drude, P. (1900) *Ann. der Physik*, **2**, 566.
2. Fermi, E. (1928) *Zeit. fur Physik*, **48**, 73.
3. Ziman, J. M. (1963) *Electrons in Metals – A Short Guide to the Fermi Surface*, Taylor and Francis, London.
4. Mackintosh, A. (1963) *Sci. Am.* **209**, 110.
5. Cracknell, A. P. and Wong, K. C. (1973) *The Fermi Surface*, Clarendon Press, Oxford.
6. Shockley, W. (1937) *Phys. Rev.*, **52**, 866.
7. Pippard, A. B. (1957) *Phil. Trans. Roy. Soc.*, **A250**, 325.
8. Pippard, A. B. (1960) *Rep. Prog. Phys.* **23**, 176.
9. Ashcroft, N. W. and Mermin, N. D. (1976) *Solid State Physics*, Holt, Rinehart & Winston, New York, p. 76.
10. De Haas, W. J. and van Alphen, P. M. (1930) *Proc. Neth. Roy. Acad. Sci.*, **33**, 1106.
11. Onsager, L. (1952) *Phil. Mag.*, **43**, 1006.

FURTHER READING

Chambers, R. G. (1990) *Electrons in Metals and Semiconductors*, Chapman and Hall, London.
Dugdale, J. S. (1976) *The Electrical Properties of Metals and Alloys*, Edward Arnold and Sons, London.
Lehmann, G. and Ziesche, P. (1990) *Electronic Properties of Metals*, Elsevier, Amsterdam.
MacDonald, D. K. C. (1956) Electrical conductivity of metals and alloys at low temperatures in *Handbook of Physics*, ed. S. Flügge, Springer, Berlin.
Moruzzi, V. L., Janak, J. F. and Williams, A. R. (1978) *Calculated Electronic Properties of Metals*, Pergamon, Oxford.
Mott, N. F. and Davis, E. A. (1971) *Electronic Processes in Non Crystalline Materials*, Clarendon Press, Oxford.
Mott, N. F. and Jones, H. (1936) *The Theory of the Properties of Metals and Alloys*, Oxford University Press.

EXERCISES

Exercise 6.1   Brillouin zones in a two-dimensional lattice. Make a plot of the first two Brillouin zones of a rectangular two-dimensional lattice with unit vectors along the x and y directions of $a = 0.2$ nm, and $b = 0.4$ nm. Give its dimensions in m^{-1}; calculate the radius of the free electron Fermi sphere if the atom has valence 1; draw this sphere on the first Brillouin zone; and show the electron band structure for both the first and second energy bands, assuming there is a small gap at the zone boundary.

Exercise 6.2   Number of k-states in reciprocal space. Show that the number of different k-states in the reciprocal space of a simple cubic lattice is equal to the number of lattice sites.

Exercise 6.3   Fermi energy of sodium and aluminium. Assuming that the free electron model applies, calculate the Fermi energy of body-centred cubic Na and face-centred cubic Al. The dimensions of the cubic unit cells in the crystal lattices are 0.43 nm and 0.40 nm respectively.

7

Electronic Properties of Semiconductors

This chapter discusses the electron band structure of semiconductors and shows how the occupancy of the electron energy levels in these materials is fundamentally different from that in metals. The reason for this is that in semiconductors and insulators in their lowest energy state, the electron bands are either filled or empty. This means that it is very difficult for the electrons to move under the action of an electric field because it would result in an increase in energy, and such energy states are not immediately available. Hence the conductivity is low. Although the energy of the electrons does vary with the wave vector k, a more simplified band structure representation is often used for semiconductors. This is the 'flat band' approach which merely represents the allowed energy levels without reference to the corresponding values of k. This approach, which is adequate for most purposes, is used widely here and in subsequent discussion of the electronic structure of semiconductors. One aspect which the flat-band model does not represent however is the difference between direct and indirect band gap semiconductors. The direct band gap materials, in which the top of the valence band and the bottom of the conduction band are located at the same point in k-space, are very important in optoelectronic applications.

7.1   ELECTRON BAND STRUCTURES OF SEMICONDUCTORS

How does the electron band structure of a semiconductor differ from that of a metal?

The electronic band structures of semiconductors and insulators are fundamentally different from those of metals because of the existence of the band gap which lies between a filled valence band and an empty conduction band. The electronic theory of semiconductors was first worked out by Wilson [1, 2]. The initial concept was that the semiconductors and insulators have highest occupied electron states which are localized in the ionic potentials (as shown in Fig. 5.1) and therefore cannot contribute to electrical conduction throughout the material. A better concept is simply that even if the top-most occupied band

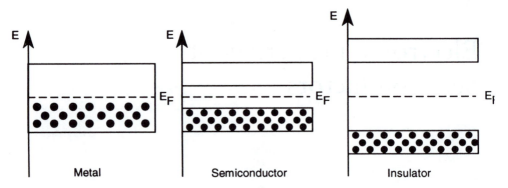

Fig. 7.1 Simplified band structure diagrams of a metal, semiconductor and insulator. Typical values of the band gap 0 eV in metals, 0.5–5.0 eV in semiconductors and 5.0 eV or greater in insulators.

Table 7.1   Typical band gaps of semiconductors

Material	Band gap E_g (eV) at 0 K	Direct/indirect gap
Ge	0.75	Indirect
Si	1.17	Indirect
GaAs	1.5	Direct
GaP	2.32	Indirect
GaSb	0.81	Direct
InSb	0.23	Direct
SiC	3.0	Indirect

$1\,eV = 1.602 \times 10^{-19}$ joules; visible light range 2–3 eV

extends throughout the material the electrons cannot conduct if it is completely filled. This definition of semi-conductors and insulators is more exact than the earlier definition given in section 5.1.5, and we will use it as a basis for further discussion.

In very simple terms then the band structures of metals, semiconductors, and insulators can be represented as shown in Fig. 7.1. Typical values of the band gap in various semiconductors and insulators are shown in Table 7.1.

7.1.1   Band structure diagrams

What band structure representation is used to interpret and predict the electronic properties of semiconductors?

Many of the electronic properties of semiconductors can be described by reference to the above simplified energy band diagrams. In fact you will find most

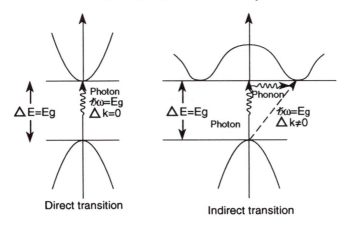

Fig. 7.2 Direct and indirect band gaps in a semiconductor.

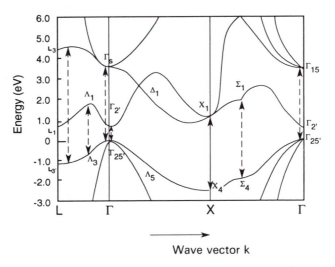

Fig. 7.3 Band structure diagram of germanium. Reproduced with permission from D. Brust, *Phys. Rev.*, **A134**, 1964, p. 1337.

textbooks base their whole discussion on these diagrams. However, the true band structure diagrams (energy versus wave vector plots) are much more complicated [3, 4, 5].

7.1.2   Direct and indirect band gaps

What do we mean by direct and indirect band gap semiconductors?

In some cases the top of the valence band and the bottom of the conduction band of a semiconductor lie at different points in *k*-space. This is called an

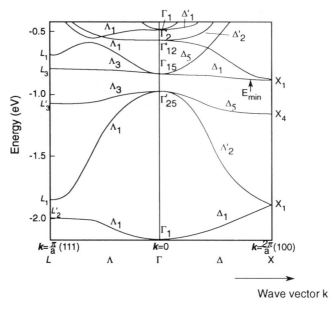

Fig. 7.4 Band structure diagram of silicon.

indirect band gap. In other cases they lie at the same point in k-space. This is called a direct band gap. These two cases are shown schematically in Fig. 7.2. The presence of a direct band gap has important consequences for optical applications of a semiconductor as we shall see later, because the probability of electronic transitions across the band gap is higher in materials with a direct band gap.

Since we have already learned how to interpret these diagrams in the case of metals, we shall look at the band structures of germanium, silicon, and gallium arsenide in k-space. These are shown in Figs 7.3, 7.4, and 7.5. In these diagrams, the band gaps are clearly shown and have the values 0.7 eV in germanium, 1.1 eV in silicon and 1.5 eV in gallium arsenide. Notice that both germanium and silicon have an indirect band gap, whereas gallium arsenide has a direct band gap.

In transforming from this type of band structure diagram to the simplified flat-band structure diagrams the connected energy levels above the gap are represented as one continuous band while the connected energy levels below the gap are represented as one continuous band. However, all information about their locations in k-space is lost as part of the simplification.

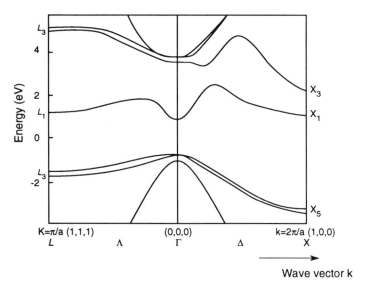

Fig. 7.5 Band structure diagram of gallium arsenide.

7.1.3 Position of Fermi level in semiconductors

Where does the Fermi level in semiconductors lie relative to the conduction and valence bands?

In a semiconductor there are no partially filled bands, just a filled valence band and an empty conduction band. Therefore the Fermi level, which separates the filled from the empty states, lies in the band gap. Consequently semiconductors do not have a well defined Fermi surface, in fact we can argue that they do not have a Fermi surface in any meaningful sense.

7.1.4 Variation of electron bands with interatomic spacing

If the interatomic spacing is changed what happens to the electron band structures?

As the interatomic spacing decreases so the electron energy bands broaden. This means that the band gap in a semiconductor should be reduced under hydrostatic pressure. This is found to occur in practice. So for example germanium, which is known to be a semiconductor under normal conditions, becomes a metal under 12 GPa (120 kbar) hydrostatic pressure.

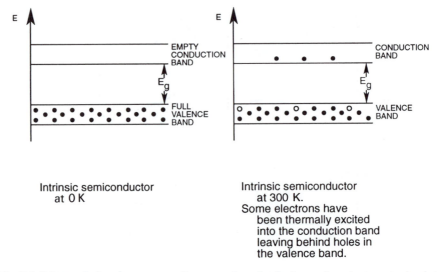

Fig. 7.6 Schematic band structure diagram of an intrinsic semiconductor at absolute zero (0 K), when the conduction band is empty, and at 300 K when some electrons have been thermally stimulated into the conduction band. The numbers in the conduction band depend on the size of the band gap and the temperature.

7.2   INTRINSIC SEMICONDUCTORS

What is an intrinsic semiconductor?

Intrinsic semiconductors are those materials with relatively small band gaps of typically less than 0.5 eV. In these cases, a number of electrons can be thermally stimulated across the band gap at room temperature (300 K), into the conduction band as shown in Fig. 7.6. Once in the conduction band these electrons contribute to the electrical conductivity, as do the 'holes' which are left behind in the valence band.

7.2.1   Thermal excitation of electrons into the conduction band

How is the number of electrons in the conduction band affected by temperature?

The electrical properties of intrinsic semiconductors are not greatly affected by the presence of impurities, at least not at room temperature. This is because the number of electrons in the conduction band is determined principally by thermal excitation of electrons from the valence band as a result of the narrow band gap.

Table 7.2 Probabilities of an electron being excited across a 1 eV band gap at various temperatures. The material will have typically 10^{28}–10^{29} electrons per unit volume in the valence band

$T\ (K)$	$k_B T\ (eV)$	$E_g/2k_B T$	$\exp(-E_g/2k_B T)$
0	0	∞	0
100	0.0086	58	0.06×10^{-24}
200	0.0172	29	0.25×10^{-12}
300	0.0258	19.3	$4.0\ \ \times 10^{-9}$
400	0.0344	14.5	$0.5\ \ \times 10^{-6}$

The Fermi distribution function $f(E)$ is given by

$$f(E) = \frac{1}{1 + \exp\left(\dfrac{E - E_F}{k_B T}\right)}.$$

If we take the band gap E_g to be typically 1 eV with the Fermi level in the middle of the gap, and the ambient temperature to be 300 K, with $k_B = 1.38 \times 10^{-23}\,\mathrm{J\,K^{-1}}$, and consequently $k_B T = 0.05\,\mathrm{eV}$, we can make the approximation,

$$f(E) \simeq \exp(-(E - E_F)/k_B T) = \exp(-E_g/2k_B T)$$

where $E - E_F \simeq E_g/2$.

This gives us the probability of an electron being thermally stimulated from the top of the valence band to the bottom of the conduction band. The values of this probability are given in Table 7.2.

Whereas $f(E)$ gives the probability of any state being occupied, the number of electrons at any given energy level, $N(E)$ is the product of the density of available electron states $D(E)$ and the probability of occupancy $f(E)$

$$N(E) = 2D(E)f(E)$$

where the factor of two is introduced because electrons can have spin up or spin down. This doubles the number of electrons that can occupy any energy level without violating the Pauli exclusion principle. The density of states $D(E)$ for free electrons as shown earlier in section 4.4.7 is given by,

$$N_0(E) = \frac{V}{6\pi^2}\left(\frac{2m}{\hbar^2}\right)^{3/2} E^{3/2}$$

which is the number of energy levels which exist below an energy E. If this expression is differentiated it gives the number of energy states within a unit energy interval $D(E)$,

$$\frac{dN_0(E)}{dE} = D(E) = \frac{V}{4\pi^2}\left(\frac{2m}{\hbar^2}\right)^{3/2} E^{1/2}$$

while the density of electronic states is twice this number, and the density of occupied states is given by

$$N(E) = \frac{V}{2\pi^2}\left(\frac{2m}{\hbar^2}\right)^{3/2} E^{1/2} f(E).$$

At room temperature an intrinsic semiconductor has about 10^{15}–10^{20} conduction electrons per m^3 (depending on the size of the band gap) caused by thermal stimulation alone. This contribution to the electrical conductivity is known as the 'dark current' simply because it arises in the absence of incident light.

7.2.2   Conductivity of intrinsic semiconductors

Which factors determine the conductivity of intrinsic semiconductors?

In the intrinsic semiconductors electrons are stimulated across the energy gap from the valence band to the conduction band. The higher the temperature, the more electrons are found in the conduction band. Each electron that undergoes such a transition leaves behind a hole in the valence band. Both electron and hole can contribute to the electrical conductivity of the material.

The conductivity σ is given by the sum of contributions due to electrons and holes,

$$\sigma = eN_e\mu_e + eN_h\mu_h$$

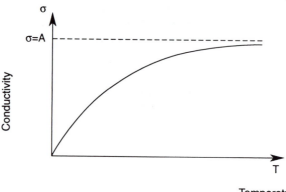

Fig. 7.7 Variation of the conductivity of an intrinsic semiconductor as a function of temperature.

where e is the electronic charge, N_e is the number density of electrons, N_h is the number density of holes, μ_e is the mobility of electrons and μ_h is the mobility of holes. The mobility is the average drift velocity of charge carriers per unit electric field strength

$$\mu = \frac{v}{\xi} = \frac{e\tau}{m} = \frac{e}{\gamma}.$$

If we assume that $N(E) = 2D(E)f(E)$ then using the above free-electron expression for $D(E)$, it can be shown that the number density of electrons and holes is

$$N_e = 4.82 \times 10^{21} \left(\frac{m^*}{m_0}\right)^{3/2} T^{3/2} \exp\left(\frac{-E_g}{2k_B T}\right)$$

$$= N_h.$$

Typically $N_e = N_h = 10^{15} - 10^{19}$ carriers per m^3 for an intrinsic semiconductor at room temperature. The electrical conductivity of an intrinsic semiconductor with band gap E_g is therefore

$$\sigma = 4.82 \times 10^{21} \left(\frac{m^*}{m_0}\right)^{3/2} T^{3/2} e(\mu_e + \mu_h) \exp\left(\frac{-E_g}{2k_B T}\right)$$

and m^* is the effective mass of the electrons and holes, which is assumed to be the same for these purposes. In fact this is rarely the case. This equation has the form

$$\sigma = A \exp\left(\frac{-E_g}{2k_B T}\right).$$

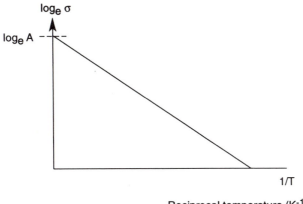

Fig. 7.8 Variation of $\log_e \sigma$ with $1/T$ for an intrinsic semiconductor.

If we take logarithms of both sides,

$$\log_e \sigma = \log_e A - \frac{E_g}{2k_B T}$$

which means that if intrinsic conductivity is the only mechanism taking place, we would expect the conductivity σ to vary with temperature t as shown in Fig. 7.8.

7.3   EXTRINSIC (OR IMPURITY) SEMICONDUCTORS

What is an extrinsic semiconductor?

An extrinsic semiconductor is a material with a large band gap which would be an insulator, except that certain defect sites are deliberately introduced which lead to additional electron or hole states (or both) in the band gap. The electrons can be thermally stimulated from the 'donor' levels in the band gap into the conduction band leading to electron conduction, or alternatively electrons can be thermally stimulated from the valence band into the 'acceptor' levels in the band gap leading to hole conduction in the valence band (Fig. 7.9).

The electrical properties of extrinsic semiconductors can be carefully controlled by the addition of acceptor or donor atoms. This means that the materials can be designed for specific technological applications, and therefore the extrinsic semiconductors remain the most important materials for electronics applications [6].

The donor levels reside at an energy ΔE_d below the conduction band and the acceptor levels reside at an energy ΔE_a above the valence band. Typically

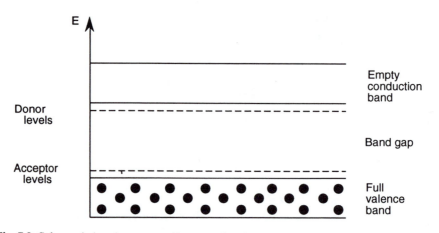

Fig. 7.9 Schematic band structure diagram of an impurity semiconductor, showing both acceptor and donor sites in the band gap.

these energy differences are about 1% of the energy gap

$$\Delta E_a \approx \Delta E_d \approx 0.01 E_g.$$

7.3.1 Donor elements

How can the number of electrons available for conduction in the conduction band be altered by the addition of impurities?

Typical donor elements in impurity semiconductors are phosphorus, arsenic, and antimony. When added in small amounts (a few parts in a million) they contribute an extra electron which populates the conduction band. The conduction mechanism in this case is via electrons, and the material is called an 'n-type' semiconductor.

7.3.2 Acceptor elements

How can the number of holes available for conduction in the valence band be altered by the addition of impurities?

Typical acceptor elements in impurity semiconductors are boron, aluminium, gallium, and indium. These have one electron less than silicon or germanium. They therefore take an electron from the valence band leaving a 'hole', which enables conduction to take place in the valence band through effective migration of these 'holes'. Since the charge carrier in this case appears to be positive, such semiconductors are known as 'p-type' semiconductors.

7.3.3 Number density and type of charge carriers

How can the number density and type of charge carriers be determined?

Unlike metals, semiconductors can have either positive or negative charge carriers. The two standard measurements which are used to determine the number and type of charge carriers are the electrical resistivity (or conductivity) and the Hall effect. We will now consider how the electronic structure of the semiconductors determines each of these properties.

7.3.4 Temperature dependence of electrical properties

How do the electrical properties of extrinsic semiconductors change with temperature?

Since the populations of electrons in the conduction band, and of holes in the valence band, increase with temperature, so the electrical conductivity of extrinsic semiconductors increases with temperature. The presence of impurities

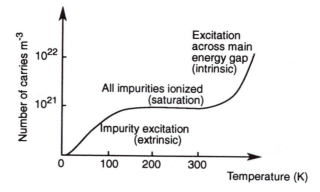

Fig. 7.10 Variation of conductivity of an extrinsic semiconductor with temperature.

provides acceptor and/or donor levels which increase the population levels of charge carriers in the valence band and conduction band respectively. Therefore, the electrical conductivity increases with impurity content (Fig. 7.10). There is also a contribution to the conductivity arising from absorption of photons, the so-called photoconductivity, provided that the photon energy is large enough to excite electrons into the conduction band from the valence band.

7.3.5   Conductivity of extrinsic semiconductors

What factors determine the conductivity of extrinsic semiconductors?

In extrinsic semiconductors we usually have a predominance of donor impurities, leading to an n-type semiconductor or a predominance of acceptor impurities, leading to a p-type semiconductor. Therefore in n-type semiconductors,

$$\sigma = N_e \mu_e e$$

and in p-type semiconductors

$$\sigma = N_h \mu_h e.$$

Typically the number density of charge carriers in an extrinsic semiconductor is $N = 10^{21} \, \text{m}^{-3}$ at room temperature.

The extrinsic or impurity contribution is for most of these semiconductors the only significant component at room temperature. At higher temperatures thermal stimulation of electrons directly across the band gap may also occur. So there are two contributions to the conductivity: intrinsic and extrinsic conduction whose relative contributions to the total conductivity are dependent on temperature.

If we plot $\log_e \sigma$ against $1/T$ as we did above for the intrinsic semiconductor, then for the extrinsic semiconductor we obtain the type of plot shown in Fig. 7.11.

Conductivity

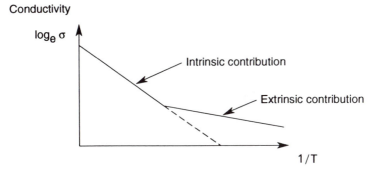

Fig. 7.11 Variation of $\log_e \sigma$ with $1/T$ for an extrinsic semiconductor.

We see from this that the extrinsic contribution is more important at low temperatures (high $1/T$) and the intrinsic contribution is more important at high temperatures (low $1/T$).

7.4 OPTICAL PROPERTIES OF SEMICONDUCTORS

How are the optical properties of semiconductors determined by the electron band structure?

Due to the band gap energy E_g, semiconductors are unable to absorb and reflect lower frequencies. Absorption and reflection start to occur at a frequency ω_0 given by $\hbar\omega_0 = E_g$ and at this frequency we notice the so-called absorption edge. At higher frequencies absorption and reflection can be relatively high. The absorption edge of a semiconductor is shown in Fig. 7.12.

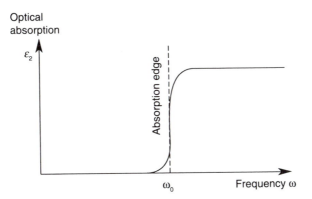

Fig. 7.12 Idealized variation of optical reflectance or absorption in a semiconductor with frequency of incident light.

7.5 PHOTOCONDUCTIVITY

What happens when electrons are excited across the band gap by absorption of photons?

In addition to thermal stimulation of electrons into the conduction band, there are other methods of achieving the same result. Illumination of a semiconductor with light of sufficient frequency (i.e. photon energy) also leads to increased numbers of charge carriers in the conduction band and hence to increased conductivity. This arises from the excitation of electrons from the valence band to the conduction band by photons. The dependence of conductivity on the absorption of light is known as photoconductivity.

The frequency of light necessary to increase the conductivity of an intrinsic semiconductor is determined by the band gap energy E_g. In an extrinsic semiconductor it is determined by the energy displacement of the donor or acceptor sites from the conduction or valence bands, respectively. The variation of conductivity with frequency of incident radiation is shown in Fig. 7.13.

For an intrinsic semiconductor with band gap 0.7 eV, the frequency ω_0 necessary to stimulate electrons across the band gap E_g is,

$$\hbar\omega_0 = E_g$$

$$\omega_0 = 1.06 \times 10^{15}\,\text{s}^{-1}$$

which corresponds to a wavelength of 1770 nm which is well into the infra-red region of the electromagnetic spectrum. Therefore optical wavelengths carry

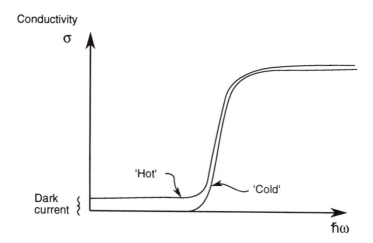

Fig. 7.13 Idealized variation of electrical conductivity in a semiconductor with frequency of incident light at constant intensity and at different temperatures.

sufficient energy to stimulate electrons across the gap in these intrinsic semi-conductors.

The dependence of photocurrent on the intensity of incident light above the threshold frequency ω_0, means that the change in electrical conductivity of a semiconductor with incident photon flux can be used as a method of detecting infra-red radiation.

The component of conduction due solely to the thermodynamic temperature is known as the 'dark current', since this is the current which would be obtained if the semiconductor was shielded from all incoming radiation (i.e. was literally kept in the dark). The variation of conductivity with frequency of light at different temperatures is also shown in Fig. 7.13. It is known that if the charge carriers are prevented from recombining immediately the photoconductivity will persist even after the incident light has been removed [7].

Extrinsic semiconductors can be used for infra-red detection provided that the frequency ω of incident infra-red radiation satisfies the condition,

$$\hbar\omega \geqslant \Delta E$$

where ΔE is the energy difference between the donor sites and the conduction band for example. In this case the threshold frequency is $\omega_0 \approx 10^{13}\,\text{s}^{-1}$. This involves photoconductivity by stimulation of electrons from donor levels or to acceptor levels. In the case of infra-red detection the semiconductor is usually cooled to 4.2 K to return thermally excited carriers to their donor or acceptor sites and so reduce the dark current, thereby improving the signal-to-noise ratio of the photocurrent.

7.6 THE HALL EFFECT

How do the electrons and holes in a semiconductor behave under the combined action of an electric field and a magnetic field?

Conductivity measurements alone are not sufficient to find the total number of charge carriers, their signs and their mobilities. Measurement of the Hall effect gives the necessary additional information.

When a current flows in a conductor or semiconductor and a magnetic field is applied perpendicular to the current, then a voltage is generated across the material in a direction perpendicular to planes containing the current and the magnetic field. This voltage is known as the Hall voltage and the phenomenon is known as the Hall effect [8].

The explanation of this phenomenon is quite simple. It arises as a result of the Lorentz force on a moving charge in a magnetic field. If a charge e is moving with velocity $\boldsymbol{v}$ in a magnetic field $\boldsymbol{H}$, then the Lorentz force $\boldsymbol{F}_\text{L}$ is given by

$$\boldsymbol{F}_\text{L} = \mu_0 e\boldsymbol{v} \times \boldsymbol{H}.$$

If we wish to express this as an equivalent electric field ξ_Hall, then remembering

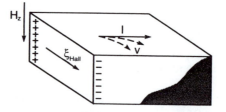

Fig. 7.14 Hall emf resulting from the action of an external magnetic field on a charge in motion. *H* is the magnetic field, *I* the conventional current, ξ_{Hall} is the Hall field, and *v* is the velocity of electrons.

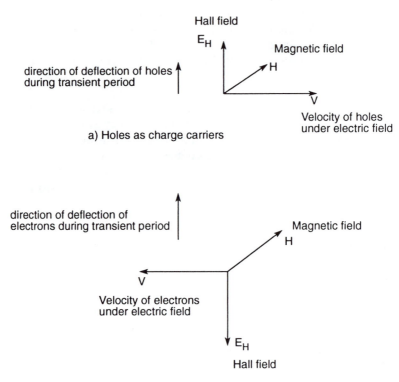

Fig. 7.15 Direction of Hall field for charge carriers of different sign. In both cases the direction of the magnetic field *H* is into the plane of the paper, and in both cases the conventional current is from left to right.

that $F_L = e\xi_{Hall}$, we have simply

$$\xi_{Hall} = \mu_0 \boldsymbol{v} \times \boldsymbol{H}.$$

We also know that the current density *J* is given by the product of *N* the number density of charge carriers, *e* the electronic charge and *v* the velocity of

Table 7.3 Hall coefficients for various materials

Material	R_H ($m^3 C^{-1}$)
Li	-1.7×10^{-10}
In	$+1.59 \times 10^{-10}$
Sb	-1.98×10^{-9}
Bi	-5.4×10^{-7}

the charge carriers

$$J = Nev.$$

Substituting this into the equation for the Hall field leaves

$$\xi_{Hall} = \mu_0 \frac{J}{Ne} \times H.$$

The Hall coefficient R_H is defined such that

$$R_H = \frac{\xi_{Hall}}{\mu_0 JH}$$

$$= \frac{1}{Ne}.$$

This means that we can make measurements of ξ_{Hall}, J and H and from these determine the product Ne. Typical values of the Hall coefficient are $R_H \simeq 10^{-9} \, m^3$ per coulomb (note that the sign depends on the sign of the charge carriers).

We note that if the Hall field ξ_{Hall} is in a certain direction for a flow of negative charge carriers, then it will be in the opposite direction for the same current when it is produced by a flow of positive charges in the reverse direction.

The Hall coefficient is inversely proportional to the number density of charge carriers N. Hall measurements are easy to make on semiconductors because N is relatively low, being typically 10^{15}–$10^{21} \, m^{-3}$. In metals, N is large being typically $10^{28} \, m^{-3}$ and the measurement of R_H is consequently more difficult. Typical values are given in Table 7.3.

The Hall effect itself can also exhibit quantum effects as shown by, among others, von Klitzing *et al.* [9].

7.7 EFFECTIVE MASS AND MOBILITY OF CHARGE CARRIERS

If the relationship between energy and k-vector in a semiconductor is no longer quadratic, how can this be expressed in terms of mobility or effective mass of the charge carriers?

We have stated in section 5.4.5 that an effective mass can be defined for electrons in a particular location in the band structure, in accordance with the relation,

$$m^* = \frac{\hbar^2}{(\mathrm{d}^2E/\mathrm{d}k^2)}$$

which means that energy bands with high curvature lead to low effective mass, and flat energy bands lead to high effective mass. For an electron near the top of a band $\mathrm{d}^2E/\mathrm{d}k^2$ is negative. This means that such an electron decelerates in the presence of a field as it exchanges momentum with the lattice.

The idea of a negative effective mass of an electron is conceptually difficult. In fact we find that the charge carriers near the top of a band are not electrons but 'holes' with a positive charge and a positive mass. The net acceleration produced is the same as that on a negative charge with a negative mass. The 'hole' moves in the opposite direction from the electronic current.

The conduction mechanism in semiconductors is therefore more complicated than in metals. Two types of charge carrier are possible, and in addition the number of charge carriers is dependent on temperature.

We have defined the mobility in section 7.2.2 as the velocity per unit field strength, $\mu = \boldsymbol{v}/\boldsymbol{\xi}$. Remember that inside the semiconductor the electrons or holes cannot accelerate indefinitely under the action of a field (as they would in free space) because of their interaction with the rest of the solid. The mobility is proportional to the inverse of the scattering probability, and we know that scattering of electrons is caused by phonons and impurity atoms. In metals the scattering probability remains almost constant as a function of temperature, but in semiconductors this is not true because the energy distribution of the carriers in a semiconductor varies with temperature.

7.8   SEMICONDUCTOR JUNCTIONS

Are there any interesting electronic effects at the boundary between two different semiconducting materials?

So far we have talked at length about the conductivity mechanisms in intrinsic and extrinsic semiconductors. We have not yet looked at the electronic properties of junctions between semiconductors, and yet from the applications viewpoint semiconductor junctions are of crucial importance in devices.

Suppose a piece of p-type semiconductor is in direct contact with a piece of n-type semiconductor. This junction has some very interesting properties which have had a direct bearing on the development of semiconductor technology. In order to understand these properties, however, we need to consider the following two principles.

1. When two solids are in contact, charge transfer occurs until their Fermi energies are the same.
2. In n- and p-type semiconductors, the Fermi level lies approximately at the donor and acceptor levels, respectively.

When the n-type and p-type semiconductors are placed in contact, an unstable situation arises temporarily because of the step change in electron and hole densities across the interface. Equalization of the Fermi levels occurs as electrons diffuse from the n-type material into the p-type material. This charges the p-type material negatively and sets up an electric field which opposes further diffusion of electrons. A dynamic equilibrium arises when the Fermi levels on either side

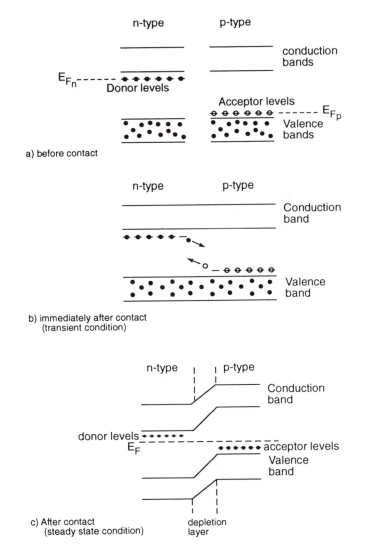

Fig. 7.16 Electron band structure diagram for a p-n junction: (a) energy levels of p-type and n-type semiconductor before contact; (b) transient condition immediately after contact; and (c) steady-state condition after contact.

are equal. However as a result of the net electrical charge, the conduction and valence bands of the two sides of the junction are displaced relative to each other, as shown in Fig. 7.16.

7.8.1 Depletion layer

How is the region in the vicinity of a semiconductor junction different from the bulk of the material?

As the electrons diffuse from the n-type to the p-type material, the number of charge carriers in the n-type material decreases, at least in the volume close to the junction. A similar effect occurs as the holes diffuse from the p-type to the n-type material. This means that in the vicinity of the interface there is a 'depletion layer' on each side of the interface containing fixed but opposite charges and a reduced number of charge carriers. This depletion layer is sometimes called the space charge region.

Now if we consider the electrical properties of this junction, it is found to have some useful properties. For example, electrons in the conduction band of the n-type material which may be trying to diffuse into the p-type material encounter a potential barrier at the junction. This makes it difficult for electrons to pass from the n-type to the p-type material, but relatively easy for them to pass in the opposite direction. This is represented in the figure by the energy gradient for electrons passing from n-type to p-type.

On the other hand, there are only a few electrons in the conduction band of

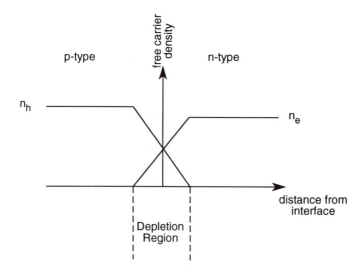

Fig. 7.17 Electron and hole densities on either side of a p-n junction.

the p-type material whereas there are many in the conduction band of the n-type material. Under equilibrium conditions, the diffusion rate of electrons in each direction is equal. Similar arguments apply to the holes in the valence band, whereby the p-type material has many holes in the valence band but the n-type material has only a few holes in its valence band.

7.8.2   Forward biasing the p-n junction

What happens if a voltage is applied across the junction to reduce the potential difference across the junction?

If we place an electric voltage across the junction, for example by connecting the positive terminal of a DC power supply to the p-type material and the negative terminal to the n-type, the voltage difference, and hence the energy difference between the two conduction bands is lowered. This is called positive or forward biasing. To a first approximation, the height of the potential barrier across the conduction band is reduced from E_g to $\Delta E = E_g - eV$ where V is the applied voltage, as shown in Fig. 7.18. As a result of this shift, more of the n-type conduction band is exposed to the p-type material and electrons cross more easily from the n-type material into the p-type material. These can then recombine with holes, emitting light. The forward-biased p-n junction can therefore be used as a light source known as a light-emitting diode, or under certain special conditions as a semiconductor laser.

7.8.3   Reverse biasing the p-n junction

What happens if a voltage is applied to increase the potential difference across the junction?

Conversely, if the negative terminal of the DC power supply is connected to the p-type material, the energy separation of the bands on either side of the junction is increased. This process is called negative or reverse biasing. In this case, because the height of the barrier is increased by $\Delta E = E_g + eV$, as shown in Fig. 7.18, it becomes very difficult to drive electrons from the n-type into the p-type material. However electrons can move easily from the p-type to the n-type material down the potential gradient of the conduction band in the vicinity of the interface. This results in an electrical current pulse and can occur for example if electrons are stimulated across the band gap by the absorption of photons in the junction itself. This means that the reverse-biased p-n junction can be used as a light detector (photodetector or photocell).

The current in a p-n junction diode therefore behaves non-linearly with the bias voltage as shown in Fig. 7.19. We see also that the simple p-n junction which we have described can act as a rectifier or diode by allowing current to flow in one direction only.

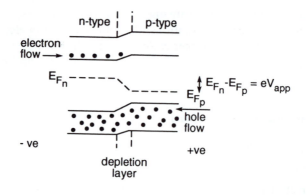

Forward bias

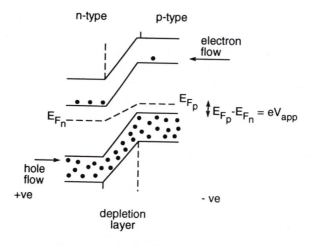

Reverse bias

Fig. 7.18 Electron band structure of reverse-biased and forward-biased p-n junctions.

7.8.4 Semiconductor devices

How can we explain the device characteristics of a p-n junction, and what other device application can be found?

We have looked at the simplest case of a semiconductor device, the p-n junction and shown that this can be used as a diode. We need to look at the operation of the p-n junction in more detail so that the current/voltage characteristics

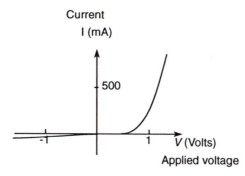

Fig. 7.19 Voltage/current characteristic of a p-n junction, showing conduction in the forward direction but no conduction in the reverse direction (until electrical breakdown occurs at much higher reverse voltages).

shown in Fig. 7.19 can be understood in terms of the electronic properties. Consider first the band structure diagram of the p-n junction shown in Fig. 7.16c.

Electrons in the conduction band of the n-type material cannot easily reach the p-type material because of the potential energy ramp which causes an internal electric field ξ. However the higher density of electrons in the conduction band of the n-type material will cause a diffusion of electrons from this region of high concentration to the region of low concentration in the conduction band of the p-type material.

In diffusion, the number of electrons passing through cross-sectional area A in unit time, dN/dt, is dependent on the rate of change of the number density of electrons with position, dN/dx, according to the equation

$$\frac{dN}{dt} = -DA\frac{dN}{dx}.$$

D is the diffusion coefficient which in this case is given by

$$D = \frac{\mu k_B T}{e}$$

where μ is the mobility of the electrons (section 7.2.2), k_B is Boltzmann's constant, e is the charge on the electron and T is the temperature in degrees Kelvin. Substituting for D in the diffusion equation

$$\left(\frac{dN}{dt}\right) = \frac{-\mu k_B T A}{e}\left(\frac{dN}{dx}\right).$$

We can therefore define a diffusion current density J_d for the charge carriers in terms of the rate of change of the number density N with time,

$$J_d = \frac{e}{A}\left(\frac{dN}{dt}\right)$$

$$= - \mu k_\mathrm{B} T \left(\frac{\mathrm{d}N}{\mathrm{d}x} \right).$$

Under equilibrium conditions this diffusion current density must be balanced by a conventional current density J_v due to the voltage gradient at the p-n junction,

$$J_\mathrm{v} = \sigma \xi = Ne\mu\xi$$

and equating the current densities, $J_\mathrm{d} = J_\mathrm{v}$

$$Ne\mu\xi = - \mu k_\mathrm{B} T \frac{\mathrm{d}N}{\mathrm{d}x}.$$

Rearranging and integrating gives,

$$V = \int \xi \mathrm{d}x = \frac{-k_\mathrm{B} T}{e} \int \frac{\mathrm{d}N}{N}$$

and noting that when $V = 0$, $N = N(0)$, and when $V = V$, $N = N(V)$, this gives

$$V = \frac{-k_\mathrm{B} T}{e} \log_e \left(\frac{N(V)}{N(0)} \right).$$

This relates the number density of electrons $N(V)$ at the top of the potential ramp of height V, (in the p-type material) to the number density of electrons $N(0)$ at the bottom of the potential ramp (in the n-type material).

If we look at the problem in more detail there is a hole current in addition to the electron current in a semiconductor. We denote the electron current density from the n-type region to the p-type region (diffusion current) as J_de, the electron current density from the p-type region to the n-type region (field current) as J_ve. Similarly the hole current densities can be denoted J_dh (diffusion) and J_vh (field).

When a bias voltage V_app is applied to the p-n junction it causes currents to flow across the junction. A positive voltage by convention reduces the potential barrier for both electrons and holes, while a negative voltage by convention increases the barrier. As the voltage is increased the net electron current density flowing from the n-type region increases. Likewise the net hole current density flowing from the p-type region increases. The current density for the diffusion of electrons can be determined from,

$$J_\mathrm{de} = \frac{e}{A} \left(\frac{\mathrm{d}N(0)}{\mathrm{d}t} \right)$$

$$= \frac{e}{A} \frac{\mathrm{d}}{\mathrm{d}t} \left(N(V) \exp \left(\frac{eV_\mathrm{app}}{k_\mathrm{B} T} \right) \right)$$

$$= \frac{e}{A} \frac{\mathrm{d}N(V)}{\mathrm{d}t} \exp \left(\frac{eV_\mathrm{app}}{k_\mathrm{B} T} \right)$$

and replacing $e(\mathrm{d}N(V)/\mathrm{d}t)/A$ by J_{ve} we arrive at,

$$J_{\mathrm{de}} = J_{\mathrm{ve}} \exp\left(\frac{eV_{\mathrm{app}}}{k_{\mathrm{B}}T}\right)$$

and similarly for the hole current

$$J_{\mathrm{dh}} = J_{\mathrm{vh}} \exp\left(\frac{eV_{\mathrm{app}}}{k_{\mathrm{B}}T}\right).$$

The net electron current density is $J_{\mathrm{e}} = J_{\mathrm{de}} - J_{\mathrm{ve}}$ and the net hole current density is $J_{\mathrm{h}} = J_{\mathrm{dh}} - J_{\mathrm{vh}}$. The total current flowing is then

$$
\begin{aligned}
J_{\mathrm{tot}} &= J_{\mathrm{e}} + J_{\mathrm{h}} \\
&= (J_{\mathrm{vh}} + J_{\mathrm{ve}})\left\{\exp\left(\frac{eV_{\mathrm{app}}}{k_{\mathrm{B}}T}\right) - 1\right\} \\
&= J_0\left\{\exp\left(\frac{eV_{\mathrm{app}}}{k_{\mathrm{B}}T}\right) - 1\right\}.
\end{aligned}
$$

This is the diode equation, where J_0 is the sum of the current densities carried by the minority carriers across the junction. We can now express the current flowing through a p-n junction in terms of the voltage across it. When $V_{\mathrm{app}} = 0$ there is no current. When $V_{\mathrm{app}} > 0$ the current increases exponentially with applied voltage. When $V_{\mathrm{app}} < 0$ the current is small and negative. This equation describes the current voltage characteristics shown in Fig. 7.19.

A vast number of semiconductor devices based on the properties of semiconductor junctions exist [10]. These include devices which are designed for particular current/voltage characteristics, such as diodes and transistors; and optoelectronic devices such as photodetectors and photoemitters. The principles of their operation remain broadly the same. We have touched on the main concepts required for understanding them in this chapter. We will look at some of these semiconductor devices in more detail when we deal with specific applications in Chapters 11–15.

7.8.5 Gallium arsenide

What is the reason for the great interest in gallium arsenide?

Although the present day semiconductor industry is based largely on silicon and to a much lesser extent on germanium, there has been widespread interest in gallium arsenide (GaAs). This has been a long-standing interest, and it is still not clear yet whether gallium arsenide will in fact fulfill its promise and revolutionize the applications of semiconductor devices. Nevertheless, we must devote some time to this material to explain why it has attracted such attention. Gallium arsenide is of great technical interest for three reasons. It has:

1. a high electron mobility μ_e
2. a large band gap E_g
3. a direct band gap.

The large electron mobility results from the small effective mass m^* of the electrons in this material. This is caused by the relatively large upward curvature of the conduction band near to the Γ point in the Brillouin zone. As a result devices can be fabricated from this material which can operate over a very short time period. The material will therefore find applications where the speed of operation of electronic systems is of paramount importance.

The large band gap of 1.5 eV makes gallium arsenide sensitive to wavelengths of light of 827 nm and shorter. This means that electrons dropping from the bottom of the conduction band to the top of the valence band will emit a photon in the red end of the visible spectrum. It can therefore be used in light-emitting diodes (LEDs) which emit red light. By combining it with other materials such as gallium phosphide, other colours can be produced by engineering the size of the band gap.

The optical properties of GaAs, specifically its high efficiency in detection and generation of light, derive from its direct band gap. This means that it can be used in fabricating lasers [11] and in optical communication devices for computers. In this application, information is transmitted by photons instead of by electrons. These optical computers are much faster than conventional computers. Recent developments in optical computing have again brought this technology to the fore.

7.8.6 Summary of gallium arsenide properties

What are the advantages of gallium arsenide over silicon and germanium for fabrication of devices?

1. GaAs circuits are faster and operate at equal, or lower, power than silicon circuits.
2. The separation between the conduction and valence bands is more easily controlled in GaAs than in silicon.
3. GaAs can radiate and detect near infra-red and visible red radiation depending on its band gap.
4. GaAs can support optoelectronic functions while silicon cannot.

REFERENCES

1. Wilson, A. H. (1931a) *Proc. Roy. Soc.*, **A133**, 458.
2. Wilson, A. H. (1931b) *Proc. Roy. Soc.*, **A134**, 277.
3. Kleinmann, W. and Phillips, J. C. (1960) *Phys. Rev.*, **118**, 1164.
4. Brust, D. *et al.* (1964) *Phys. Rev.*, **134**, 1337.

5. Pollak, M. *et al.* (1966) *Physics of Semiconductor Devices*, Kyoto.
6. Osburn, C. M. and Reisman, A. (1987) *J. Electr. Mater.*, **16**, 223.
7. Lowney, J. R. and Mayo, S. (1992) *J. Electr. Mater.*, **21**, 731.
8. Hall, E. H. (1879) *Amer. J. Math.*, **2**, 287.
9. Von Klitzing, K., Dorda, G. and Pepper, M. (1980) *Phys. Rev. Letts.*, **45**, 494.
10. Morant, J. D. (1970) *Introduction to Semiconductor Devices*, 2nd edn, Harrap, London.
11. Goodhue, W. D. *et al.* (1990) *J. Electr. Mater.*, **19**, 463.

FURTHER READING

Leaver, K. D. (1989) *Microelectronic Devices*, Longmans, Harlow, England.
Solymar, L. and Walsh, D. (1984) *Lectures on the Electrical Properties of Materials*, 3rd edn, Oxford University Press.
Sze, S. M. (1981) *Physics of Semiconductor Devices*, 2nd edn, John Wiley & Sons, New York.
Van der Ziel, A. (1968) *Solid State Physical Electronics*, 2nd edn, Prentice Hall, New Jersey.

EXERCISES

Exercise 7.1   Approximation to the Fermi function in semiconductors. If the band gap in a semiconductor is $E_g = 0.5$ eV and the temperature is 300 K, show that the Fermi function can be approximated by,

$$f(E_g) = \exp(-E_g/2k_B T)$$

and that in general the probability of an electron being found in the conduction band at an energy ΔE above the Fermi level is

$$f(E) = \exp(-\Delta E/k_B T).$$

Exercise 7.2   Temperature dependence of conductivity in intrinsic semiconductors. A sample of Ge exhibits intrinsic conductivity at 300 K. If the absorption edge is at 1771 nm, estimate the temperature rise that will result in a 30% increase in conductivity.

Exericse 7.3   Electronic properties of gallium arsenide, silicon and germanium. Compare the known electronic properties (e.g. band gap, electron mobility, conductivity, optical properties, etc.) of silicon, germanium and gallium arsenide. Indicate engineering applications where GaAs has a distinct advantage over the others. If the $E - k$ relationship for the bottom of the conduction band of a specimen of GaAs is of the form $E = Ak^2$, where $A = 7.5 \times 10^{-38}$ Jm^2, find the effective mass of the conduction electrons in kilograms, and as a ratio of the free-electron rest mass.

8

Electrical and Thermal Properties of Materials

In previous chapters we have built up an understanding of the electronic properties of materials on the microscopic scale. These theories have involved first considering the electrons as classical particles of a free-electron gas and later as free-electron waves contained within the material. We then found that the electrons occupy allowed energy bands and we were able to distinguish between on the one hand metals, and on the other semiconductors and insulators on the basis of their electron band structures. We also found that these electron bands were anisotropic and so plots of allowed energy against position in k-space were necessary. Now we must bring all of these ideas together to account for the macroscopic electrical and thermal properties of materials.

In this chapter therefore we look at the relationship between macroscopic measurable electrical properties and the underlying electronic properties such as mobility, effective mass and number density of electrons. Then we look at various thermoelectric effects which span the interface between electrical and thermal effects. Finally we discuss the phenomenon of thermoluminescence which bridges the gap between thermal and optical properties of materials, and therefore provides a link to the next chapter on optical properties.

8.1 MACROSCOPIC ELECTRICAL PROPERTIES

How can the macroscopic properties of materials be described on the basis of the preceding microscopic theories?

Our objective here is to describe the macroscopic electrical properties of materials and then explain these through models of the microscopic mechanisms inside the material. The familiar macroscopic properties are the conductivity and the Hall effect.

8.1.1 Generalized Ohm's law and conductivity

Can the relation between current and voltage be explained in both metals and semiconductors on the basis of the classical free-electron theory?

Although Ohm's law is often written in the elementary form $V = IR$ where I is the electrical current flowing in a resistance R under an applied voltage V, the law can also be written in the equivalent form,

$$J = \sigma \xi$$

where J is the current density, σ is the conductivity and ξ is the electric field. As shown in section 3.2.1 the current density J can be written in terms of the number density of electrons N, their charge e and their average drift velocity v

$$J = Nev.$$

The drift velocity v can be expressed in terms of the mobility μ which is defined as the drift velocity per unit field (we have already encountered this mobility briefly in Exercise 3.3),

$$v = \mu \xi.$$

Combining these equations gives the conductivity as,

$$\sigma = Ne\mu$$

and so

$$J = Ne\mu \xi.$$

In metals the charge carriers are electrons and so we are concerned in this case only with the number density, charge and mobility of electrons. In semiconductors both electrons and holes contribute to the conduction, as discussed in section 7.2.2, so that using a similar equation for the current density in terms of both contributions from holes and electrons leads to

$$J = (N_e \mu_e + N_h \mu_h) e \xi.$$

In order for Ohm's law to hold, we see that neither N nor μ can be a function of electric field ξ. In fact, under certain conditions both N and μ can become functions of ξ and Ohm's law is no longer valid. We may, however, regard the above equation as a more generalized form of Ohm's law in which this can be taken into account.

8.1.2 Temperature dependence of conductivity in metals

Can the temperature dependence of conductivity in metals be described by the classical free-electron theory?

If we assume one type of charge carrier for simplicity, then the temperature dependence of the conductivity is dependent on the temperature dependence

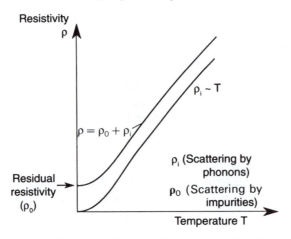

Fig. 8.1 Resistivity as a function of temperature in metals. This consists of two components, one due to impurity scattering ρ_0 which persists even at zero temperature, and one due to phonon scattering ρ_i.

of N and μ, because e is clearly constant according to our present understanding. In a metal N is the density of valence ($=$ conduction) electrons. This has a value of typically $N = 10^{28} \, \mathrm{m}^{-3}$ in a metal, and is largely temperature independent. Therefore the temperature dependence of conductivity should be due to a temperature dependence of mobility.

The mobility of electrons in metals is of the order of $\mu = 10^{-3}$–$10^{-1} \, \mathrm{m}^2 \, (\mathrm{Vs})^{-1}$ and so this leads to a conductivity σ of typically 10^6–$10^8 \, (\mathrm{ohm \, m})^{-1}$. In fact all of the observed temperature dependence of σ in metals arises from the temperature dependence of the electron mobility μ which is affected by phonon scattering and impurity scattering of electrons in the metal (Fig. 8.1). The classical free-electron theory does not give any inherent indication of how μ should vary with temperature. Therefore the temperature dependence of electrical conductivity can only be described by an *ad hoc* variation of mobility with temperature in the classical free-electron model.

8.1.3   Temperature dependence of conductivity in semiconductors

Can the temperature dependence of conductivity in semiconductors be described by the classical electron theory?

In intrinsic semiconductors, the number density of charge carriers increases with temperature according to the equation

$$N = N_0 \exp\left(\frac{-E_\mathrm{g}}{2k_\mathrm{B} T}\right)$$

where E_g is the band gap and the above equation assumes that the Fermi level is in the middle of the band gap. This equation shows that there is an increase in the number density of conduction electrons with temperature. In addition, there is a change in mobility of the charge carriers with temperature, but this is less significant than the change in charge carrier density. Therefore in semiconductors the temperature variation of N dominates the temperature dependence of conductivity.

8.1.4 Temperature dependence of mobility

How can the temperature dependence of mobility of electrons be explained?

If we return to the classical description of electrons moving in a material, their motion is continually disrupted by scattering. If the mean free time between collisions is τ, the charge e and the mass m, then the mobility μ is given by,

$$\mu = \frac{e\tau}{m} = \frac{e}{\gamma}$$

and it can be seen that it is the temperature dependence of τ which determines the mobility, or alternatively we can view this as the temperature dependence of the resistive coefficient γ in the equation of motion of the electrons. Therefore in a metal γ increases with temperature, leading to a reduction in mobility μ and so a decrease in conductivity σ.

The classical model gives no indication of the temperature dependence of γ, although it is reasonable to suppose that, as the temperature is raised, the increased vibrations of the lattice will cause more collisions with the free electrons and contribute to a higher resistive coefficient γ or shorter mean free time τ.

8.1.5 Different types of mobility

How can we define electron mobility in a material?

There are four different kinds of mobility of electrons in a material which must be distinguished.

1. Microscopic mobility

$$\mu_{\text{mic}} = \frac{v}{\xi}.$$

This is defined for a particular electron moving with drift velocity v in an electric field ξ. It therefore cannot easily be experimentally verified, and so remains only a concept from which a more practical description of collective mobility of electrons can be developed.

2. Conductivity mobility

$$\mu_{con} = \frac{e\tau}{m^*}.$$

This is the macroscopic or average mobility which is determined from measurement of electrical conductivity σ,

$$\sigma = Ne\mu_{con}$$

assuming N and e are both known.

3. Hall mobility

$$\mu_H = \sigma R_H = \frac{\sigma \xi_{Hall}}{\mu_0 JH}$$

is the mobility of charge carriers as determined from a Hall effect measurement.

4. Drift mobility

$$\mu_d = \frac{d}{\xi t}.$$

This is determined from measurement of the time t required for carriers to travel a distance d in the material under the action of an electric field ξ.

8.2 QUANTUM MECHANICAL DESCRIPTION OF CONDUCTION ELECTRON BEHAVIOUR

Do all 'conduction' electrons actually contribute to the electrical conductivity?

As we have shown in Chapter 5, electrons in a material behave not like classical particles but like waves [1]. This leads to properties which are different from classical expectations. In the absence of an electric field, the valence electrons in a metal have no net or preferential velocity in any direction. If we plot the vectors of these electrons in velocity space, then for a free-electron metal we obtain a velocity sphere, the surface of which corresponds to the Fermi velocity. All points inside the Fermi sphere are occupied. Integrating over the entire sphere we obtain zero drift velocity.

When an electric field is applied, the Fermi sphere is displaced as shown in Fig. 8.2. Still the majority of electron velocities cancel, but now some are uncompensated and it is these electrons which cause the electric current. We note therefore the important result that only certain specific electrons which are close to the Fermi surface can contribute to the conduction mechanism. Note that a similar effect was found for heat capacity where only those electrons within $k_B T$ of the Fermi level could contribute to the heat capacity.

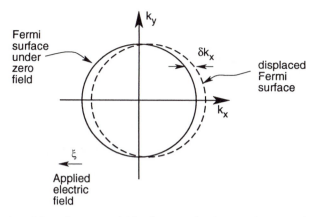

Fig. 8.2 Velocity of free electrons within the Fermi sphere under zero electric field and under an applied field ξ along the x direction.

8.2.1   Quantum corrections to the conductivity in Ohm's law

How is Ohm's law modified if only the electrons close to the Fermi surface contribute to the conductivity of a metal?

The highest energy that electrons can take in a metal in its ground state is the Fermi energy E_F. We also know that the density of occupied states is highest around E_F, since for a free-electron model the density of states $D(E)$ has the following form, as shown in section 4.4.7

$$D(E) = \frac{V}{4\pi^2}\left(\frac{2m}{\hbar^2}\right)^{3/2} E^{1/2}.$$

This means that only a small change of energy ΔE is needed to raise a large number of electrons above the Fermi level. We will consider that the velocity of the uncompensated electrons under the action of the field ξ is close to the Fermi velocity. This will be a reasonable simplifying assumption. With this in mind, we can calculate the electric conductivity σ, based on quantum mechanical considerations.

Our Ohm's law equation of section 8.1.1 needs to be slightly modified to take into account the fact that not all free electrons contribute to the conductivity. Hence,

$$J = N^* e v_F$$

where v_F is the velocity of electrons at the Fermi level and N^* is the number of displaced electrons, that is those in the shaded region of Fig. 8.3 which contribute to the conductivity.

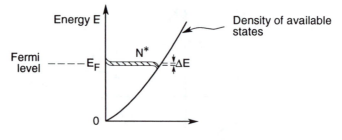

Fig. 8.3 Population density versus energy for free electrons according to the free-electron theory.

8.2.2 Number of 'conduction' electrons contributing to conduction

How can we find out how many of the so-called conduction electrons in a metal actually contribute to electrical conduction?

We need to obtain an expression for N^*. This will clearly be dependent on the density of occupied states at the Fermi level $N(E)$ and the displacement energy ΔE,

$$N^* = N(E)\Delta E$$

and consequently

$$J = N(E)\Delta E e v_{\text{F}}$$

$$= v_{\text{F}} e N(E) \frac{\mathrm{d}E}{\mathrm{d}k} \Delta k.$$

The term $(\mathrm{d}E/\mathrm{d}k)$ is determined from the energy versus wave vector diagram for the given case. For free electrons, we have $E = \hbar^2 k^2 / 2m$ and hence,

$$\frac{\mathrm{d}E}{\mathrm{d}k} = \frac{\hbar^2 k}{m} = \hbar v_{\text{F}}$$

and this yields

$$J = v_{\text{F}}^2 e N(E) \hbar \Delta k.$$

8.2.3 Displacement of the Fermi sphere under the action of an electric field

How does the displacement of the electron wave vectors depend on other factors such as the mean free time of electrons between collisions?

Now we need to find the displacement of the Fermi sphere Δk under the influence of the electric field ξ. Since we know that $m(\mathrm{d}v/\mathrm{d}t) = e\xi$, and since $p = \hbar k$ is the momentum, it follows that the force on the electrons can be expressed as

$$F = m\frac{\mathrm{d}\boldsymbol{v}}{\mathrm{d}t} = \hbar\frac{\mathrm{d}k}{\mathrm{d}t} = e\xi.$$

So,

$$\mathrm{d}k = \frac{e\xi}{\hbar}\,\mathrm{d}t$$

or

$$\Delta k = \frac{e\xi}{\hbar}\Delta t = \frac{e\xi}{\hbar}\tau$$

where τ is the mean free time of the electrons between collisions. With this expression for Δk, we arrive at the following expression for the current density

$$\boldsymbol{J} = \boldsymbol{v}_{\mathrm{F}}^2 e^2 N(E)\xi\tau.$$

Only the projections of $\boldsymbol{v}_{\mathrm{F}}$ along the direction of the electric field ξ, that is $\boldsymbol{v}_{\mathrm{F}}\cos\theta$, contribute to the current

$$\boldsymbol{J} = e^2 N(E)\xi\tau \int_{-\pi/2}^{+\pi/2} (\boldsymbol{v}_{\mathrm{F}}\cos\theta)^2\,\frac{\mathrm{d}\theta}{\pi}$$

$$= \tfrac{1}{2}e^2 N(E)\xi\tau\boldsymbol{v}_{\mathrm{F}}^2.$$

For a spherical Fermi surface there is a slight correction which gives,

$$\boldsymbol{J} = \tfrac{1}{3}e^2 N(E)\xi\tau\boldsymbol{v}_{\mathrm{F}}^2$$

and finally, the conductivity is given by $\sigma = \boldsymbol{J}/\xi$, so that

$$\sigma = \tfrac{1}{3}e^2 \boldsymbol{v}_{\mathrm{F}}^2\tau N(E).$$

This quantum mechanical statement of conductivity shows that not all 'conduction' electrons can contribute to the conductivity, but only those close to the Fermi surface. In addition, the conductivity is determined by the density of occupied states near the Fermi level. For metals such as copper, which has one conduction electron per atom, this density is high, leading to high conductivity. For bivalent metals such as calcium which has two conduction electrons per atom, this density is small, leading to a relatively low conductivity. Therefore it is the density of states at the Fermi surface, and not the classically expected total number of conduction electrons, which determines the conductivity of a material.

8.3 DIELECTRIC PROPERTIES

How can we represent the response of a non-conducting material to an electric field?

Most electronic applications involve the use of alternating electric fields or currents. In these cases the atoms in insulators oscillate under the action of the applied electric field, and these oscillations can be expressed in terms of the dielectric constant, ε. This is often expressed in terms of real and imaginary components ε_1 and ε_2

$$\varepsilon = \varepsilon_0 \varepsilon_r = \varepsilon_1 + i\varepsilon_2.$$

This dielectric 'constant' is actually dependent on the frequency of the applied electric field. When considering its dependence on the frequency of electromagnetic radiation it is often represented as $\varepsilon(\omega)$.

8.3.1 Polarization

How do we quantify the displacement of charge in a material under the action of an electric field?

The net result of applying an electric field to an insulator is to cause the positive and negative charges within the material to become displaced in opposite directions and the material consequently becomes electrically polarized. The polarization can result from the relative displacement of the electrons and ionic cores or alternatively from the relative displacement of positive and negative ionic cores [2].

The force F on a charge e under the action of an electric field is,

$$F = e\xi$$

and it is this force which causes polarization of a material by displacing the positive and negative charges within an atom in opposite directions, or by displacing the ionic cores within the lattice.

The electric polarization of the material is denoted by the symbol P. This is an electric dipole moment per unit volume, which is measured in coulomb metres per cubic metre (or effectively coulombs per square metre). The equation for P is

$$P = Np$$

where p is the dipole moment of an individual atom and N is the number of atoms per unit volume. P can also be defined as the surface density of charge which appears on the faces of the specimen when placed in a field. The polarization can be expressed in terms of the electric field by the equation,

$$P = \varepsilon\xi = \varepsilon_r \varepsilon_0 \xi$$

where ε is the permittivity or dielectric constant. We see therefore that the dielectric constant is a measure of the amount of electric polarization induced by unit field strength. A high dielectric constant means that a material is easily polarized in an electric field. Typical values of the relative dielectric permittivity

ε_r are in the range 1.0–10 (dimensionless), although its value can be much higher in some special materials, for example ε_r is 94 in titanium dioxide (TiO_2) [3. p. 267].

8.3.2 Dielectric field strength

How high an electric field can a material withstand before it suffers electrical breakdown?

The dielectric field strength is a measure of the largest electric field strength that an insulating material can sustain before the electrostatic forces holding the atoms in place are overcome. Once this happens the material suffers electrical breakdown and suddenly becomes an electrical conductor. This breakdown voltage has been discussed in Chapter 1. Typical values of the dielectric strength are in the range of megavolts per metre. However it should be noted that the breakdown strength often increases with frequency, and in particular for most materials breakdown is somewhat inhibited above 10^8 Hz [3. p. 272].

8.3.3 Electrical properties of noncrystalline materials

What about materials that do not have a regular crystalline lattice?

We have looked in detail at the electrical properties of crystalline materials. In these cases, the regular periodicity of the atoms on the lattice sites leads to relative simplicity of calculation. However, we need not restrict ourselves entirely to these materials since electrical conduction also occurs in polymers, ceramics, and amorphous materials and there is a need to provide theories and models for these materials also.

8.3.4 Polymers

Why are some polymers found to be good conductors?

Most polymers are insulators of course, but conducting polymers exist which have electrical properties resembling those of conventional metals or semiconductors [4]. Polyacetylene contains a high degree of crystallinity and a relatively high conductivity compared with other polymers. *Trans*-polyacetylene has a conductivity that is comparable to silicon. The electron band structure of this polymer has even been calculated and it has been found that when all carbon lengths are equal, this material has a band structure which is reminiscent of a metal. When the carbon bonds alternate in length it is found that band gaps appear in the structure.

8.4   OTHER EFFECTS CAUSED BY ELECTRIC FIELDS, MAGNETIC FIELDS AND THERMAL GRADIENTS

What other effects occur when a material is subjected to external influences such as electric, magnetic and thermal fields?

A number of other phenomena occur when a material is subjected to electric, magnetic or thermal fields. We will mention only the most important of these: magnetoresistance, the Seebeck effect, the Peltier effect, the Nernst effect and the Ettingshausen effect.

8.4.1   Magnetoresistance

What happens to the electrical resistance when a material is subjected to a magnetic field?

The magnetoresistance is the change in electrical conductivity associated with an applied magnetic field. It cannot be explained by the classical (Drude) electron model since with one carrier conductivity, constant relaxation time τ and constant effective mass the magnetoresistance is zero.

An example of zero magnetoresistance occurs in our explanation of the Hall effect in which the deflection of charge carriers causes the build-up of a transverse electric field which exactly counteracts the effect of the magnetic field. In this case under equilibrium conditions the motion of the charge carriers is identical in the presence or absence of a magnetic field, because of this transverse electric field, resulting in zero magnetoresistance.

However if not all the charge carriers have the same properties, the current flow is disturbed by the presence of a magnetic field and some of the charge carriers travel a longer distance between electrodes than in the absence of a field. This leads to a larger observed resistivity and the difference between the zero field resistivity and the measured resistivity under the applied magnetic field known as magnetoresistance.

The resistivity is defined as $\rho = \xi/J$ where ξ and J, the electric field and current density, are measured along the same direction. Since the resistivity is in general dependent on the magnetic field we find that,

$$\rho = \xi/J = \rho_0(1 + \omega_c^2\tau)$$

where $\omega_c = eB_\perp/m$ and $B_\perp$ is the magnetic induction perpendicular to the direction of measurement of current,

$$\rho = \rho_0\left(1 + \frac{e^2B_\perp^2}{m^2}\tau\right)$$

and τ is the mean free time of the electrons between collisions. The magnetoresistive term is then simply

$$\rho_{\text{mag}} = \rho_0 \frac{e^2 B^2}{m^2} \tau.$$

8.4.2 Thermoelectric power (Seebeck effect)

What happens to the voltage across a material when it is subjected to a temperature field (temperature gradient)?

If a material is subjected to a temperature gradient, the energy of the carriers at the hot end is greater than at the cold end and this leads to a carrier concentration gradient along the material. Displaced charge resulting from this concentration gradient generates a counteracting electric field ξ until the total current becomes zero. The magnitude of this electric field in terms of the voltage per degree difference is known as the thermoelectric power α. In a metal,

$$\alpha = \frac{dV}{dT} \simeq \frac{\pi^2 k_{\text{B}}^2 T}{e E_{\text{F}}}$$

where E_{F} is the Fermi energy. In a metal α is typically a few microvolts per degree Kelvin.

In a semiconductor, for an n-type material

$$\alpha = -\frac{k_{\text{B}}}{e} \left(A + \frac{E_{\text{c}} - E_{\text{F}}}{k_{\text{B}} T} \right)$$

and for a p-type material

$$\alpha = -\frac{k_{\text{B}}}{e} \left(A + \frac{E_{\text{F}} - E_{\text{v}}}{k_{\text{B}} T} \right).$$

Here A is a constant which depends on the specific scattering mechanism, $A = 2$ for lattice scattering and $A = 4$ for charged impurity scattering, E_{c} is the energy level at the bottom of the conduction band, E_{v} is the energy at the top of the valence band and E_{F} is the Fermi energy. In a semiconductor α is typically a few millivolts per degree Kelvin.

The Seebeck effect is utilized in the thermocouple which is used for measuring temperature. The thermoelectric power α is determined from the open circuit electric field ξ caused by a temperature gradient dT/dx

$$\alpha = \frac{\xi}{(dT/dx)}.$$

8.4.3   Peltier effect

What happens to the temperature gradient when a current flows in a material?

When a current flows in a material, a temperature gradient is developed. This of course is the inverse of the Seebeck effect and is used in some cases for temperature control. The Peltier coefficient π is simply the ratio of the electrical current density J to the thermal current density J_Q

$$\pi = \frac{J_Q}{J}.$$

8.4.4   Nernst effect

What happens when both a magnetic field and a thermal field (temperature gradient) are applied simultaneously to a material?

When a magnetic field is applied at right angles to a temperature gradient, the diffusing charge carriers are deflected in the same way as when the magnetic field is applied at right angles to a conventional electric current. The result is a Nernst voltage. However, since charge carriers of both signs diffuse in the same direction the polarity of the Nernst voltage is not dependent on the sign of the charge carrier.

8.4.5   Ettingshausen effect

Do we therefore also get a transverse thermal field (temperature gradient) in the Hall effect?

In the Hall effect, the application of a magnetic field normal to the passage of an electric current leads not only to a transverse voltage but also to a transverse temperature gradient. The appearance of this temperature gradient is known as the Ettingshausen effect. This arises because charge carriers with different energies (velocities) are deflected differently by the magnetic field. This is a small effect which adds to the Hall voltage.

8.5   THERMAL PROPERTIES OF MATERIALS

Which factors determine the thermal properties of materials?

The thermal properties of materials can be determined principally by the electrons, as in the case of thermal conductivity of metals, or principally by the lattice, as in the case of thermal conductivity of insulators or of specific heat capacity [5].

8.5.1 Thermal conductivity

How does thermal conduction take place in materials?

Thermal conductivity of materials varies from $6 \times 10^3 \, \text{W m}^{-1} \text{K}^{-1}$ in silver and copper to $5 \times 10^{-2} \, \text{W m}^{-1} \text{K}^{-1}$ in sulphur [6]. In the case of metals, the thermal conduction mechanism is similar to the electrical conduction mechansim and proceeds via the free electrons which migrate throughout the material. In semi-conductors, conduction can take place by the electrons which are thermally stimulated into the conduction band.

In insulators another mechanism must be involved and in this case the thermal conduction is due to phonons which are created at the hot part of a solid and destroyed at the cold part. These phonons provide the mechanism by which energy is transferred through the material. In metals the phonon contribution to thermal conductivity is also present, but the electronic contribution is so much greater that in these cases the phonon contribution is neglected.

We have already defined the thermal conductivity K in section 1.5.1. It is the ratio of the heat flux J_Q ($= Q/A$) to the thermal gradient dT/dx

$$K = - \frac{J_Q}{(dT/dx)}$$

and its units are $\text{W}^{-1} \text{m}^{-1} \text{K}^{-1}$.

8.5.2 Mechanism of thermal conduction

How can we develop a theory of thermal conduction based on our knowledge of the electronic properties?

If we begin from the assumption that thermal conduction can arise from both the motion of free electrons and phonons, we can derive a theory of the thermal conductivity. Again as in electrical conductivity, only those electrons close to the Fermi surface can contribute to the thermal conductivity.

8.5.3 Thermal conductivity of metals

How is heat conducted in metals?

From quantum mechanics we have shown that in the electrical conduction process, only those electrons close to the Fermi surface can absorb energy and hence contribute to the conductivity. The same is also true for the thermal conductivity. Therefore, to a very good approximation, the velocity of those electrons contributing to the thermal conductivity is the velocity at the Fermi level v_F.

The number of participating electrons N^* is determined by the population density at the Fermi energy $N(E_F)$. To a first approximation, this is about 1% of the number of free electrons per unit volume. As we have already seen in section 4.5.1 this gives the following contribution to the electronic heat capacity

$$C_v^e = \frac{\pi^2}{2} N k_B \frac{T}{T_F}$$

$$C_v^e = \frac{\pi^2}{2} \frac{N k_B^2 T}{E_F}.$$

We will now show that a relationship exists between the specific heat capacity and the thermal conductivity.

We consider the situation depicted in Fig. 8.4 where l is the mean free path length of the electrons between collisions. Assume that heat flow is linear only along the x direction and is zero in the plane perpendicular to the x direction. In this case let us consider a section of length $2l$ with unit cross-sectional area. The heat flux is

$$J_Q = E_{out} - E_{in}$$

where E_{in} is the heat energy flowing in per unit time per unit area at the left end, and E_{out} is the heat energy flowing out per unit time per unit area at the right end. This can be written as,

$$J_Q = 2z\frac{3}{2}k_B l \frac{dT}{dx}$$

where z, the number of electrons per unit time per unit area impinging on the end face, is

$$z = \frac{1}{6} N^* v.$$

We might reasonably assume that N^*, the number density of free electrons contributing to thermal conductivity, is similar to the number contributing to

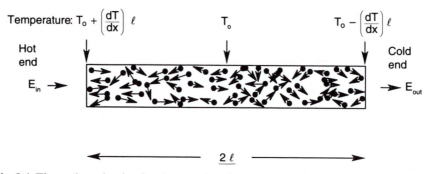

Temperature: $T_o + \left(\frac{dT}{dx}\right)\ell$ T_o $T_o - \left(\frac{dT}{dx}\right)\ell$

Hot end E_{in} →   → E_{out} Cold end

2ℓ

Fig. 8.4 Thermal conduction by electrons in a linear section of material under the action of a thermal gradient.

the thermal heat capacity $N^* \simeq 2C_v^e/3k_B$, since in both cases the electrons must be able to absorb heat energy. Substituting for z,

$$J_Q = -\frac{N^*}{2}vk_Bl\frac{dT}{dx} = \frac{1}{3}C_v^e vl\frac{dT}{dx}$$

and from the earlier equation for the thermal flux $J_Q = K(dT/dx)$, it is clear that

$$K = \frac{1}{3}C_v^e vl.$$

Here l is the mean path length between two electron collisions and v is the average velocity of electrons. Notice the important result that the thermal conductivity and electronic heat capacity are related. This is at first a surprising result.

Since only electrons close to the Fermi level can take part in thermal conductivity, we will replace N^* with N_F the number density of electrons at the Fermi surface, replace l with l_F the mean free path of electrons at the Fermi level, and replace v with v_F the velocity of electrons at the Fermi surface. Substituting into the above expression for C_v^e and using the more precise quantum mechanical expression for heat capacity $C_v^e = (\pi^2/2)(N_F k_B^2 T/E_F)$ from section 4.5.1, the equation for K becomes

$$K = \frac{\pi^2}{6E_F}N_F k_B^2 T v_F l_F.$$

Therefore the thermal conductivity increases with mean free path l_F, number of electrons per unit volume at the Fermi surface N_F, and velocity of electrons at the Fermi surface v_F. Remembering that $E_F = m^* v_F^2/2$, and that $l_F = \tau v_F$

$$K = \frac{\pi^2}{3m^*}N_F k_B^2 T\tau.$$

The thermal conductivity increases with T and τ and decreases with m^*. We can reasonably have expected the conductivity to have increased with the density of states at the Fermi level and the mean free time between electron collisions. Similarly we expect conductivity to increase with increasing electron mobility (or decreasing effective mass). However we might at first be surprised at the increase with temperature. This however is taken care of in the temperature dependence of τ, which can be used to compensate for this.

8.5.4 Thermal conductivity of insulators

How are thermal conductivity and heat capacity related in insulators?

Once again the thermal conductivity K is related to the heat capacity by the expression

$$K = \tfrac{1}{3}C_v^l vl$$

but now C_v^1 is the lattice heat capacity of the phonons, v is the phonon velocity and l is the phonon mean free path.

8.6   OTHER THERMAL PROPERTIES

What other thermal properties are of interest?

We now go on to consider other thermal properties which have important but less wide-ranging application. One of these is thermoluminescence.

8.6.1   Thermoluminescence

When semiconductors or insulators are heated is there any emission of light?

Thermoluminescence is the emission of electromagnetic radiation, in the visible spectrum, when certain materials are heated [7]. These materials must be either insulators or semiconductors, and they must have a large number of electrons trapped in impurity states in the band gap.

The emitted radiation from thermoluminescent materials is different from the well-known black-body radiation (incandescence) which depends on the fourth power of the absolute temperature, Stefan's law. A typical thermoluminescence 'glow curve' is shown in Fig. 8.5.

8.6.2   Mechanism of thermoluminescence

What distinguishes thermoluminescence from incandescence in terms of the electronic properties?

If we have an electron band structure in which there is a band gap, with a number of isolated defects or impurity states in the band gap, and a Fermi

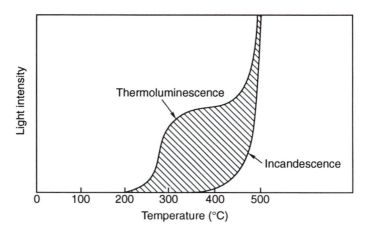

Fig. 8.5 Thermoluminescence glow curve of emitted light intensity versus temperature.

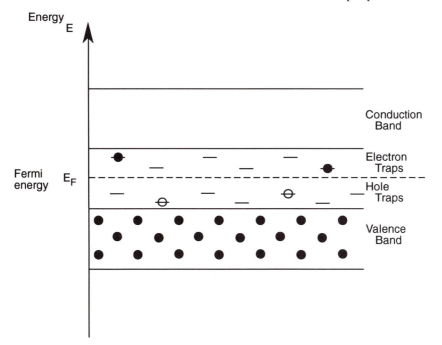

Fig. 8.6 Schematic band structure diagram showing band gap with electron and hole traps. Charged particles are held in the traps for long periods. When they escape, they recombine and emit light.

level between the bands as shown in Fig. 8.6, then electrons can be trapped in these impurity states.

Electrons become trapped in these localized energy levels by being stimulated into the conduction band and then dropping down into the localized energy states in the band gap, instead of back into the valence band as illustrated in Fig. 8.7.

The lifetime of electrons in the traps depends on a number of factors, including the prevailing temperature T and the depth of the trap below the conduction band. The lifetime can actually be many years, and this is made use of in thermoluminescent dating of pottery and other ceramics for example, and in radiation dose monitoring using thermoluminescent sensors (section 16.5). The requirements for a material to be able to exhibit thermoluminescence are:

(a) presence of a band gap;
(b) presence of impurity energy states in the band gap;
(d) long lifetime of electrons in traps;
(d) material must have been subjected to radiation to excite electrons from valence band before becoming trapped;
(e) material must not have been inadvertently heated, which could empty electrons from traps.

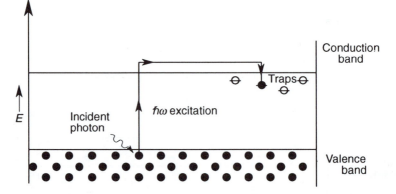

Fig. 8.7 Mechanism of electron trapping in thermoluminescence: (1) electron is excited into the conduction band; (2) electron moves freely within the conduction band; and (3) electron falls into localized energy state (electron trap).

8.6.3   Theory of thermoluminescence

How is the light emitted in thermoluminescence?

Once we have electrons located in traps in the band gap we need to explain how this leads to the emission of light. Essentially electrons are thermally stimulated from the traps into the conduction band and later they fall back into the valence band, emitting a photon as they do so, as in Fig. 8.8.

If we have an electron located in a trapped state at an energy ΔE below the conduction band, then the probability of the electron being thermally stimulated

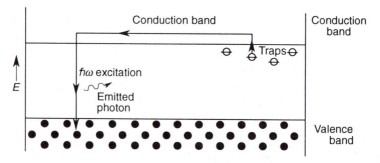

Fig. 8.8 Mechanism of electron excitation and thermoluminescent emission of light: (1) thermal stimulation from trap to conduction band; (2) movement within conduction band; and (3) transition to valence band with photon emission.

into the conduction band in unit time is given by the Arrhenius equation,

$$p = s \exp\left(\frac{-\Delta E}{k_B T}\right)$$

where s is a constant, with dimensions time^{-1} and typically of magnitude 10^{11}–$10^{17} s^{-1}$. This means that there is a time frame associated with the occupancy of the electron trap once the electron is there.

8.6.4 Occupation and vacation of trapped states by electrons

How does the occupancy of electron traps vary with time?

The probability of filling any state in the band gap will also be dependent on time. If dN/dt is the rate of stimulation of electrons from traps into the conduction band, then

$$\frac{dN}{dt} = -Np$$

where N is the number of electrons in traps and p the probability of escape in unit time. This simply states that the number of events leading to stimulation of electrons into the conduction band is proportional to the number of electrons sitting in traps. Integrating this equation gives,

$$-\log_e\left(\frac{N}{N_0}\right) = \int_0^t p \, dt$$

$$N = N_0 \exp(-pt)$$

and substituting for p,

$$N = N_0 \exp\left(-s \exp\left(\frac{-\Delta E}{k_B T}\right)t\right)$$

This is the Randall–Wilkins equation [8] which describes the number of electrons remaining in traps as a function of both time t and temperature T. We know that eventually an electron must escape from a trap, and so the integral of p over the time interval $t = 0$ to $t = \infty$ must be unity

$$\int_{t=0}^{\infty} p \, dt = 1.$$

8.6.5 Lifetime of electrons in traps

How can we determine the time an electron will, on average, stay in the electron trap?

Clearly the lifetime of occupancy of an electron state is inversely proportional to the probability p of a transition in unit time. We may define this lifetime τ

as a function of temperature T by

$$\tau(T) = \frac{1}{p(T)}$$

$$= \frac{1}{s}\exp\left(\frac{\Delta E}{k_B T}\right).$$

Therefore raising the temperature T decreases the expected lifetime of the electrons in the traps. This is what we should expect, since more thermal energy increases the probability of the electron escaping by thermal stimulation.

From the exponential decay equation $N = N_0 \exp(-pt)$ it is possible to define a half-life for the occupancy of the electron traps. Simply, when the number of traps remaining occupied has declined to half, $N = N_0/2$, we have the half-life of the occupancy $\tau_{1/2}$

$$\tfrac{1}{2} = \exp(-p\tau_{1/2})$$

$$\tau_{1/2} = \frac{1}{p}\log_e 2.$$

8.6.6   Intensity of light emitted during thermoluminescence

What factors determine the intensity and frequency of light emitted during thermoluminescence?

The intensity of light emitted during thermoluminescence is dependent on the rate of emptying of the electron traps dN/dt. If we assume that every electron from a trap enters the bottom of the conduction band and then instantaneously falls back to the top of the valence band with emission of a photon of energy equal to the band gap energy, then light of a single frequency $v = (E_g/h)$ will be emitted. The intensity of the light will be equal to the rate of emptying of electron traps,

$$I = -\frac{dN}{dt}$$

$$= Np$$

$$= Ns\exp\left(-\frac{\Delta E}{k_B T}\right)$$

where ΔE is the energy difference between the traps and the conduction band.

8.6.7 Emission of light on heating

How does the emitted light intensity depend on time and temperature for a single type of electron trap?

Suppose then the temperature of the specimen is raised at a constant rate,

$$\frac{dT}{dt} = \beta$$

then the fractional change in occupancy (dN/N) is

$$\frac{dN}{N} = -p \, dt$$

$$\frac{dN}{N} = -s \exp\left(\frac{-\Delta E}{k_B T}\right) dt.$$

Replacing dt with dT/β gives

$$\frac{dN}{N} = -\frac{s}{\beta} \exp\left(\frac{-\Delta E}{k_B T}\right) dT.$$

Integrating this expression to give the number of occupied states leads to the following expression,

$$\log_e\left(\frac{N}{N_0}\right) = -\int_{T_1}^{T_2} \frac{s}{\beta} \exp\left(\frac{-\Delta E}{k_B T}\right) dT$$

$$N = N_0 \exp\left\{ -\int_{T_1}^{T_2} \frac{s}{\beta} \exp\left(\frac{-\Delta E}{k_B T}\right) dT \right\}$$

and since we have stated that the intensity of radiation is given by $I = Ns \exp(-\Delta E/k_B T)$ we are led to the conclusion

$$I = N_0 s \exp\left\{ -\int_{T_1}^{T_2} \frac{s}{\beta} \exp\left(\frac{-\Delta E}{k_B T}\right) dT \right\} \exp\left(\frac{-\Delta E}{k_B T}\right).$$

This emission assumes a single type of trap at an energy ΔE below the conduction band, a constant rate of change of temperature and a constant value of s for all traps of the given type.

8.6.8 Location of the peaks in thermoluminescent intensity

How can the depths of the electron traps below the conduction band be studied from the thermoluminescent glow curve?

When intensity of emission I is measured as a function of temperature T as the temperature is swept at a fixed rate, peaks in the intensity will correspond to the depth of electron traps below the conduction band.

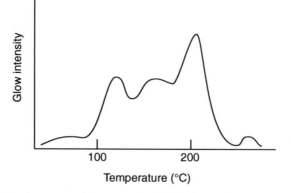

Fig. 8.9 Thermoluminescence glow curve for TLD 100 dosimeter with several intensity peaks corresponding to several depths of electron traps. Reprinted from *Nucl. Tracks and Radiation Meas.*, **11**, R. K. Bull, p. 108, copyright 1986, with kind permission from Pergamon Press Ltd, Oxford, UK.

An empirical relationship has been given between the depth ΔE in electron volts (eV) and the peak temperature T^* in Kelvin (K) by Urbach [9]

$$\Delta E \text{ (eV)} = \frac{T^*}{500} \text{ (K)}.$$

8.6.9   Applications of thermoluminescence

How is thermoluminescence used in its primary applications of radiation dosimetry and archaelogical dating?

Thermoluminescence finds applications in radiation dosimetry [10], geological, and cosmological dating and in the investigation of radiation damage in solids. The thermoluminescent glow curve gives information about the total radiation dose absorbed by the material. For dosimetry this is all that is required.

The elapsed time since formation of a solid can be calculated for dating purposes assuming a certain average background radiation intensity over a period of time, and assuming the material has not been heated in the interim. If the material has been heated this would empty some or all of the electron traps, effectively resetting the thermoluminescent clock.

A useful review of thermoluminescence and its applications has been given by Bull [11]. Thermoluminescent radiation detectors are discussed in section 16.5.

REFERENCES

1. Brillouin, L. (1953) *Wave Propagation in Periodic Structure*, Dover Press, New York.
2. Anderson, J. C. (1964) *Dielectrics*, Chapman and Hall, London.
3. Solymar, L. and Walsh, D. (1984) *Lectures on the Electrical Properties of Materials*, Oxford University Press.

4. Mort, J. and Pfister, G. (1982) *Electronic Properties of Polymers*, J. Wiley & Sons, New York.
5. Kittel, C. and Kroemer, H. (1980) *Thermal Physics*, 2nd edn, W. H. Freeman, San Francisco.
6. Tye, R. P. (1969) *Thermal Conductivity*, Academic Press.
7. Garlick, G. F. J. and Gibson, A. F. (1984) *Proc. Phys. Soc.*, **A60**, 574.
8. Randall, J. T. and Wilkins, M. H. F. (1945) *Proc. Roy. Soc.*, **A184**, 366.
9. Urbach, F. (1930) *Wiener Berichte*, **IIA**, 139, 363.
10. McKinlay, A. F. (1981) *Thermoluminescence Dosimetry*, Adam Hilger, Bristol.
11. Bull, R. K. (1986) *Nucl. Tracks Radiation Meas.*, **11**, 105.

FURTHER READING

Anderson, J. C., Leaver, K. D., Rawlings, R. D. and Alexander, J. M. (1990) *Materials Science*, 4th edn, Chapman and Hall, London, Chapters 7 and 13.
Hummel, R. (1993) *Electronic Properties of Materials*, 2nd edn, Springer-Verlag, Berlin, Parts II and V.
Kittel, C. and Kroemer, H. (1980) *Thermal Physics*, 2nd edn, W. H. Freeman, San Francisco.
Mott, N. F. and Davis, E. A. (1971) *Electronic Processes in Non-Crystalline Materials*, Clarendon Press, Oxford.
Tye, R. P. (1969) *Thermal Conductivity*, Academic Press, New York.
Ziman, J. M. (1960) *Electrons and Phonons*, Oxford University Press.

EXERCISES

Exercise 8.1 Drift velocity of conduction electrons. The Fermi energy of aluminium is 12 eV and its electrical resistance at 300 K is $3 \times 10^{-8}\,\Omega\,\text{m}$. Calculate the mean free path of the conduction electrons and their mean drift velocity in a field of $10^3\,\text{V}\,\text{m}^{-1}$. (Atomic weight of aluminium = 27, density = 2700 kg m^{-3}).

Exercise 8.2 Conductivity in intrinsic and extrinsic semiconductors. A sample of n-type germanium contains 10^{23} ionized donors per cubic metre. Estimate the ratio at room temperature of the conductivity of this material to that of high purity intrinsic germanium. Assume the band gap in germanium is 0.7 eV.

Exercise 8.3 Thermoluminescence and lifetime of electrons in traps. If the lifetime of trapped electrons in a particular ceramic at 273 K is 10^{10} s (320 years), and if the frequency parameter $s = 4.64 \times 10^{17}\,\text{s}^{-1}$ calculate the depth of the electron traps below the conduction band. Then calculate the temperature at which the peak occurs in the thermoluminescence glow curve, and calculate the lifetime of the same electrons in the same traps at a temperature of 373 K.

9

Optical Properties of Materials

We have touched briefly on the optical properties of materials in the early chapters, but here we must bring together the concepts of electron structure and the known optical properties of materials. This is done by identifying the allowed energy transitions which determine the main features of the optical spectrum. This means that we need to connect measured optical properties with the allowed electron energy levels. The major classification of electron transitions is between transitions within the same band (intraband) and transitions between different bands (interband). The former are lower energy transitions which lead to the high reflectivity of metals in the visible spectrum. The latter are higher energy transitions which can lead to specific colours in materials. Various methods for measuring the optical properties are discussed including both conventional static optical measurements and differential techniques under external modulation of field, temperature or stress. Finally the specialized topics of photoluminescence and electroluminescence are discussed.

9.1 OPTICAL PROPERTIES

What quantities need to be measured to completely determine the optical properties of a material?

In previous chapters we have shown that the optical properties of materials can be described in terms of two constants. These are the refractive index n and the extinction coefficient k. Alternatively we can choose the real and imaginary components ε_1 and ε_2 of the dielectric 'constant' or complex permittivity. The reflectance R can be expressed in terms of either of these two pairs of parameters [1].

9.1.1 Penetration depth δ, and attenuation coefficient α

How can we describe empirically the reduction in intensity of light when it passes through a material?

When discussing the electronic transitions in materials which arise from the absorption of photons we should remember that these do not necessarily take place throughout the bulk of the specimen. The depth of penetration of incident light depends on the frequency of the light and the optical constants of the material. The depth at which the intensity of the incident electromagnetic wave is attenuated to $1/e$ of its value is called the penetration depth δ. This is expressed by the following equation,

$$I = I_0 \exp(-z/\delta)$$

where z is the distance into the material. Replacing $1/\delta$ by the attenuation coefficient α, which is also widely used to charactérize materials, gives the relation,

$$I = I_0 \exp(-\alpha z).$$

In transparent materials, such as various different types of glass, δ is large, being of the order of 0.1–0.3 m, while in metals δ is very small, being of the order of 10^{-8} m.

9.1.2 Physical significance of the optical constants n and k

How do the observed optical constants relate to the absorption of a wave in a material medium?

The solution of the wave equation in a material with optical constants n and k leads to the following equation for the electric vector ξ [2]

$$\xi_x(z) = \xi_0 \exp\left(\frac{-\omega k z}{c}\right) \exp\left(i\omega\left\{t - \frac{nz}{c}\right\}\right)$$

$$= \left(\begin{array}{c}\text{incident}\\\text{amplitude}\end{array}\right) \times \left(\begin{array}{c}\text{damping}\\\text{term}\end{array}\right) \times \left(\begin{array}{c}\text{undamped}\\\text{oscillation}\end{array}\right)$$

Here, ξ_x is the electric field component parallel to the surface, ω is the frequency of the incident radiation, z is the distance normal to the surface of the material, x is a direction parallel to the surface of the material and c is the velocity of the incident light wave. The optical constants n and k have been defined in sections 1.4.1 and 1.4.2. Since $\omega/c = 2\pi/\lambda$ this equation can be expressed alternatively in terms of the wavelength λ

$$\xi_x(z) = \xi_0 \exp\left(\frac{-2\pi k z}{\lambda}\right) \exp\left(i\left\{\omega t - \frac{2\pi n z}{\lambda}\right\}\right).$$

9.1.3 Dielectric constants of materials

How are the optical constants of a material related to the dielectric constants?

The above equation for the parallel component of ξ as a function of depth z contains two terms, an exponentially decaying term which is dependent on k

and an undamped wave term which is dependent on n. Therefore n affects the phase of the light wave in the material and k affects its amplitude. The optical properties can be expressed equally in terms of the real and imaginary parts of the dielectric constant ε_r as follows,

$$\varepsilon_r = \varepsilon_1 + i\varepsilon_2 = (n + ik)^2$$

where the total dielectric constant is $\varepsilon = \varepsilon_0(\varepsilon_1 + i\varepsilon_2)$.

The intensity of light, which is proportional to ξ^2, is then given by,

$$I = \xi^2$$

$$= I_0 \exp\left(\frac{-2\omega kz}{c}\right)$$

and from this equation we can use the definition of the penetration depth δ as the distance required to decrease the intensity by a factor of $1/e$

$$\frac{I}{I_0} = \exp\left(\frac{-2\omega k\delta}{c}\right) = \frac{1}{e}.$$

Under these conditions the penetration depth δ is,

$$\delta = \frac{c}{2\omega k} = \frac{\lambda}{4\pi k}$$

and the attenuation coefficient is

$$\alpha = 2\omega k/c = \frac{4\pi k}{\lambda}.$$

Notice that δ depends on k but not on n. Some typical values of extinction coefficient k and penetration depth δ in the visible range of the spectrum are given in Table 9.1.

We see therefore, that while optical properties of materials such as water and glass are the result of a bulk measurement, in graphite and gold they are restricted to measurements made over a few tens of nanometers at the surface. This is shown in Fig. 9.1. Once again the fact that light only penetrates a few

Table 9.1 Values of extinction coefficient (k) and penetration depth (δ) for various materials in the visible range of the spectrum

Material	k	$\delta\,(m)$
Water	1.4×10^{-7}	0.32
Glass	1.5×10^{-7}	0.29
Graphite	0.8	60×10^{-9}
Gold	3.2	15×10^{-9}

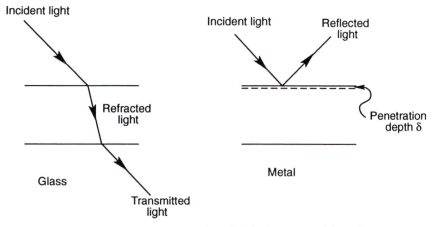

Fig. 9.1 Refraction and reflection of light by a material medium.

nanometers in some materials implies that those materials must have high reflectance. Reflectance measurements on metals are highly sensitive to the surface condition (e.g. presence of oxide coating) and a question also remains whether a surface measurement under these conditions is representative of bulk material.

9.2 INTERPRETATION OF OPTICAL PROPERTIES IN TERMS OF SIMPLIFIED ELECTRON BAND STRUCTURE

How can the features of the optical reflectance spectrum be related to the electron band structure?

We have also shown in section 6.2.1 that the optical properties can be explained in terms of the electronic properties of the materials. Simplified band structure diagrams for a metal and a semiconductor are shown in Fig. 9.2. The high reflectance of metals is a result of the partially filled conduction band which allows photons to be absorbed and reflected over a wide range of energies, forming a continuum of energies from the infra-red up to the visible range. In the visible or ultra-violet however, an energy is reached beyond which the absorption and reflection usually decrease markedly in metals as shown in Fig. 3.7 due to an inability to excite electrons to just above the top of the conduction band. The energy at which this occurs represents the energy between the Fermi level and the top of the conduction band.

In semiconductors, reflectance and absorption are low in the infra-red, but absorption becomes possible as soon as the photon energy becomes larger than the band gap. Consequently, semiconductors have higher absorption and reflection in the ultra-violet. This has been shown in Fig. 7.12.

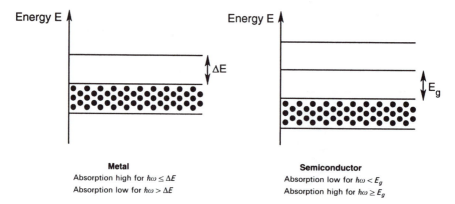

Fig. 9.2 Schematic band structure diagrams for a metal and semiconductor.

9.2.1   Summary of optical absorption processes

How can the various electronic transitions be classified?

(a) Interband transitions

(i) The highest energy transitions are those from the bottom of the valence band to the top of the conduction band. Changes in the density of states across these bands affect the absorption of light at different energies and so give rise to a frequency dependence of the attenuation coefficient.

(ii) Other lower energy interband transitions from the top of the valence band to the bottom of the conduction band also occur. The 'absorption edge' occurs at $\hbar\omega = E_g$, the gap energy. The attenuation coefficient α in semiconductors is usually in the range 10^7–$10^8 \, \text{m}^{-1}$ for energies above the band gap energy E_g. However α decreases by several orders of magnitude once the energy drops below the band gap energy E_g because there are no longer energy states for the excited electrons to occupy so they cannot absorb the energy of the incident photons.

(iii) Another electronic process is known as exciton generation. It is an excitation which produces a bound electron–hole pair. The electron is trapped in a localized energy level in the band gap while the hole remains mobile in the valence band. The exciton can dissociate into independent free carriers or can recombine with the emission of a photon or phonon.

(iv) Excitation of electrons from localized trap sites in the band gap into the conduction band can occur at energies lower than E_g. This usually occurs from optical absorption, although it can also arise from thermal excitation. The optical absorption arising from this process is much lower than for other interband transitions because there are relatively few trapped electrons compared with electrons in the valence band.

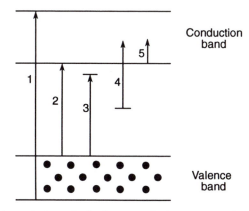

Fig. 9.3 Classification of the principal types of electron transitions: (1) high energy interband transitions; (2) transitions across the band gap–absorption edge; (3) exciton generation (bound electron–hole pair); (4) impurity level excitation; and (5) intraband transition.

(b) Intraband transitions

In metals the absorption of photons by electrons occurs over a continuous wide range of energies beginning effectively from zero energy. This usually involves the absorption or emission of phonons to conserve momentum. In this process the electrons move between energy states in the same band. The intraband transitions can only occur in metals, and they are responsible for the high reflectivity of metals at low energies.

The various types of electronic transitions are represented on the 'flat band' diagram of Fig. 9.3.

9.2.2 Colour of semiconductors

How is the colour of a material determined by its electronic structure?

The band gaps of certain semiconductors, such as the III–V semiconductors can be changed by alloying. If the band gap of a range of semiconductors varies from 3.5 eV (ultra-violet) to 1.5 eV (infra-red), then when these are illuminated with white light the colour of the materials by transmission changes as progressively more of the visible spectrum is absorbed, beginning from the high-energy blue end of the spectrum and ending with all optical energies being absorbed. The colours change from colourless to yellow, orange, red and finally black, depending on whether all of the visible spectrum or only a portion of the longer wavelength region is transmitted. Here of course certain colours by transmission are not possible because of the nature of the absorption process.

We should also note that colour by transmission and colour by reflection will be different in these cases because of the interband absorption process. So a material with a band gap in the yellow region of the visible spectrum at 2 eV might, when illuminated with white light, appear orange-red by transmission but blue-green by reflection, since only the blue-green portion of the spectrum can be absorbed and hence reflected.

Colour itself is a subjective phenomenon. For example a suitable combination of yellow and blue light may appear green to the eye, even though a spectral analysis would reveal that each of the original frequency components is still present. This is simply due to the physiology of the human eye which interprets the presence of certain frequencies of light as colour in a non-unique way. The eye detects colours only in terms of the combinations of 'primary' colours: red (565 nm), green (535 nm) and blue (445 nm). It is possible to persuade the eye that certain colours (i.e. frequencies) are present even when the actual spectrum is merely a suitable combination of these 'primary' colours. Consequently there is nothing inherently fundamental about the so-called 'primary' colours, they merely represent frequencies to which the human eye is sensitive.

9.2.3   Direct and indirect transitions across the band gap

How does the probability of an interband transition depend on the difference between the energy of the photon and the band gap?

If the probability of a direct, that is k-conserving, transition is calculated it is found to be dependent on the square root of the difference between the photon energy and the band gap energy. The attenuation α is proportional to this probability and therefore

$$\alpha = \alpha_d(\hbar\omega - E_g)^{1/2}.$$

Naturally the probability of a transition is zero when $\hbar\omega < E_g$. Consequently, a plot of α^2 against $\hbar\omega_{photon}$ gives a straight line with intercept equal to the band gap energy E_g.

The probability of indirect, that is phonon-assisted, transitions is much lower than for direct transitions. This leads to a lower value of attenuation coefficient, by typically two or three orders of magnitude. The effects of indirect interband transition on the optical properties of solids are therefore only noticeable in the absence of direct transitions. However the transition depends on $(\hbar\omega - E_g)^2$, that is

$$\alpha = \alpha_i(\hbar\omega - E_g)^2.$$

Here α_i for indirect transitions is much smaller than α_d for direct transitions.

9.2.4   Impurity level excitation

How can electrons escape from local traps in the band gap?

The elevation of a trapped electron from an impurity level (or electron trap) in the band gap to the conduction band can occur either by thermal excitation,

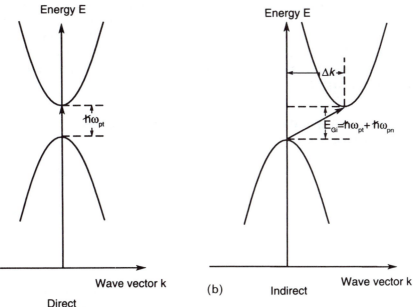

Fig. 9.4 Direct and indirect interband transitions shown on an E versus k diagram.

or alternatively by absorption of a photon. Usually, in order that thermal stimulation can occur the energy of the trap has to be close enough to the conduction band so that $\Delta E \approx k_B T$. Excitation by a photon is simply dependent on the energy of the photon being greater than the difference in energy between the trapped state and the bottom of the conduction band, $\hbar\omega \geqslant \Delta E$.

Imperfections, defects, or impurity levels are localized and so do not extend throughout the solid. Therefore they are represented as a short line on the energy band diagram. These imperfections usually are one of the following types:

- point defects
- impurities
- dislocations and grain boundaries.

9.2.5   Purity of semiconductor materials

How closely is the impurity content controlled in 'electronic quality' semiconductor materials?

In semiconductor materials in which the engineering of band structures to meet stringent requirements is essential, it is clear that the presence of unanticipated defects must be kept to an absolute minimum. Otherwise the materials will have unexpected and undesirable electronic properties. Therefore, production of semiconductor materials takes places in extremely clean environments. The purity of semiconductor materials is frequently better than one part in a million (excluding the doping materials which are also on the level of parts in a million). In metals, 'high purity' usually means something like 99.9% or one part in a thousand. We see therefore that much greater care is needed in the fabrication of semiconductors than is needed for metals.

9.2.6   Identification of the occurrence of interband transitions from band structure diagrams

How can we locate the electronic transitions from an E versus k diagram?

It is possible to interpret the optical spectra in terms of the electron band structure of a material [3]. If we take a very simple example of a metal with a band structure as shown in Fig. 9.5 below, we can identify the optical transitions that are possible.

Here the minimum separation between the bands occurs at the zone boundary. Since both bands are empty at this point, no transitions can occur here. A direct transition can occur at the Fermi surface to the next highest band with energy $\hbar\omega_0$. Transitions are then possible from all other occupied lower energy states to the corresponding points in the upper band. These are k-conserving. The highest energy transition is $\hbar\omega_m$ from the bottom of the occupied band to the top of the next unoccupied band.

Electronic interband transitions are only possible from the region of k-space from $-k_F$ to $+k_F$ in Fig. 9.5, which in this case represents the occupied states below the Fermi level. The transitions with the lowest energy occur at $k = -k_F$ from the Fermi level to the unoccupied level in the conduction band at $k = k_F$. That is to say these are direct (k-conserving) transitions

$$\Delta E_{min} = \hbar\omega_0.$$

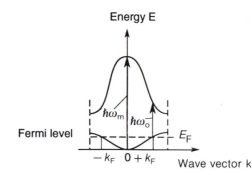

Fig. 9.5 Direct electronic transitions at different locations within the Brillouin zone.

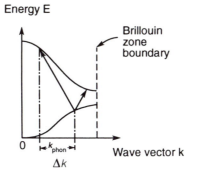

Fig. 9.6 Indirect interband transitions described within the first Brillouin zone.

Other transitions are possible. Direct transitions over the range of energies from $\hbar\omega_0$ to $\hbar\omega_m$ are possible. The highest energy direct interband transition $\hbar\omega_m$ occurs in this case at the centre of the zone, the Γ point.

Indirect, that is phonon-assisted, non-k-conserving, interband transitions are also possible, but these occur with much lower probability.

An example of an indirect interband transition is shown in Fig. 9.6. The change in momentum of the electron is $+\Delta k$, therefore a phonon of momentum $-\hbar\Delta k$ must be emitted to conserve momentum.

9.2.7 Intraband transitions

What low-energy transitions are possible in metals?

Transitions between electron states in the same band are called intraband transitions. These always need the assistance of a phonon, and so are indirect

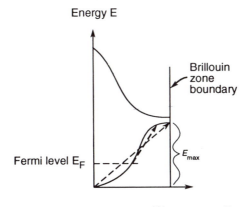

Fig. 9.7 Indirect, phonon-assisted intraband transition within the conduction band of a metal.

transitions (Fig. 9.7). They can only occur from occupied states to unoccupied states at the Fermi level and above.

These transitions are the archetypal 'free-electron' transitions which are used to describe the optical absorption and reflection processes in the classical model. These can occur in metals only, and are responsible for infra-red absorption and high reflectance in the optical frequency range.

9.3   BAND STRUCTURE DETERMINATION FROM OPTICAL SPECTRA

How can the details of the electron band structure be investigated through measurements?

The principal methods of determining the electron band structures of materials are optical methods, although a range of other techniques are used to give supplementary information, including photoemission studies, de Haas–van Alphen effect and theoretical band structure calculations [4].

Experimentally, reflectance is the easiest optical property to measure, but in most cases R is a rather slowly varying function of wavelength and this makes it very difficult to locate the exact energies of interband transitions. Also R itself does not contain all the available optical information. The absorption ε_2, or the extinction coefficient k, are much more useful since they have rather sharper features, but even these need supplementing with ε_1 or n data respectively

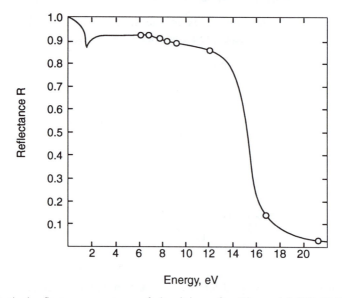

Fig. 9.8 Optical reflectance spectrum of aluminium after Ehrenreich [5]. © IEEE 1965.

to completely specify the optical properties. As an example, consider the reflectance of aluminium which is shown in Fig. 9.8.

Here we see a fairly typical metallic reflectance spectrum with high reflectance at low energy and a sharp decay at about 15 eV, much as we might expect on the basis of the Drude model. Notice some structure at about 1.5 eV, but otherwise the spectrum is featureless. The absorption ε_2 as shown in Fig. 9.9 has sharper features, so that the existence of a transition at 1.5 eV is clearly indicated.

9.3.1   Case studies: optical reflectance and band structure

What do the electron band structures of real materials look like?

The optical absorption spectrum of aluminium is shown in Fig. 9.9. The main spectral features occur at an energy of 1.5–2.0 eV. These can be attributed to parallel band absorption along the Z direction between the X and W points in the Brillouin zone, where the band separation is 2 eV over a wide region between the W point and the Fermi surface [6] as shown in Fig. 5.17. Another region of parallel band absorption is along the Σ direction between the zone centre Γ and the K point. Here the parallel bands are typically 1.5 eV apart.

The optical absorption spectrum for germanium is shown in Fig. 9.10. The main features occur at energies of 0.7, 2.3 and 4.5 eV. These correspond to the transitions $\Gamma_{25'} \rightarrow \Gamma_{2'}, \Lambda_3 \rightarrow \Lambda_1$ and $\Sigma_4 \rightarrow \Sigma_1$ as shown in the band structure diagram of Fig. 7.3. The first of these is the band gap across the Γ point which is not quite the minimum band gap energy in germanium.

Figure 9.11 shows the absorption spectrum for copper which is fairly typical of a free-electron metal with high absorption at low energies. Spectral features occur at 2, 5, 6 and 7 eV, these features correspond respectively to interband

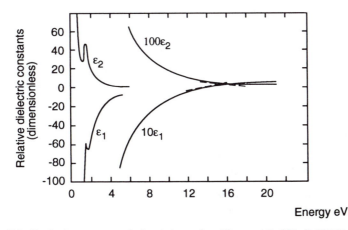

Fig. 9.9 Optical constants of aluminium after Ehrenreich [5]. © IEEE 1965.

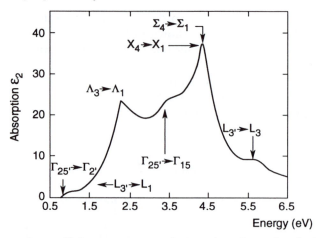

Fig. 9.10 Absorption coefficient ε_2 spectrum of germanium. Reproduced with permission of F. Abeles, *Optical Properties and Electronic Structure of Metals and Alloys*, published by Elsevier, 1966.

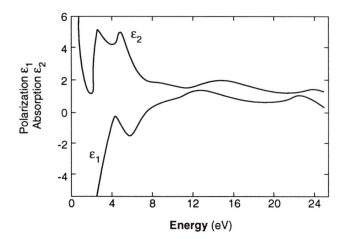

Fig. 9.11 Polarization ε_1 and absorption coefficient ε_2 spectra of copper [5]. Reproduced with permission from H. Ehrenreich and H. R. Philipp, *Phys. Rev.*, **128**, 1962, p. 1622.

transitions $Q_1 \rightarrow Q_2$ at the Fermi level, $X_5 \rightarrow X_{4'}$, $L_{2'} \rightarrow L_1$ and $\Sigma_2 \rightarrow \Sigma_3$ at the K point as shown in the band structure diagram of Fig. 5.16.

9.3.2 Modulation spectroscopy

Is it possible to accentuate the interband spectral features relative to the broad background intraband absorption?

Further enhancement of the spectra can be obtained by differentiation. This is achieved experimentally by a collection of techniques known as modulation

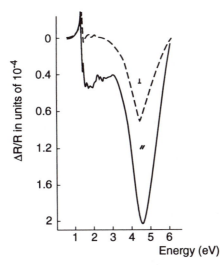

Fig. 9.12 Piezoreflectance spectrum of aluminium [7]. Reprinted from *Solid State Communications*, **47**, D. C. Jiles, p. 38, copyright 1983, with kind permission from Pergamon Press Ltd, Oxford, UK.

spectroscopy. In these methods the optical spectrum is modulated by the superposition of alternating strain (piezoreflectance) [7], temperature (thermoreflectance), electric field (electroreflectance), magnetic field (magnetoreflectance) and wavelength [8, 9].

All of these, with the exception of wavelength modulation, cause cyclic changes, or perturbations in the electron band structure. These emerge as changes in the optical properties. They cause an enhancement of the interband transitions over the intraband transitions in the optical spectrum because the intraband contribution to the optical properties is almost independent of the modulation, even though the energy levels do change; whereas the sensitivity of the band gap to modulation causes significant changes in the available interband transition energies. An example of the enhancement due to strain of the reflectance spectrum of aluminium is shown in Fig. 9.12.

The modulation spectroscopy techniques therefore lead to an enhancement of features in the optical spectra by measuring the derivative of the optical parameters R, n, and k (or R, ε_1 and ε_2) with respect to strain, electric field, temperature, or magnetic field.

9.4 PHOTOLUMINESCENCE AND ELECTROLUMINESCENCE

What other methods are there for causing emission of light from semiconductors and insulators?

In the previous chapter we discussed thermoluminescence which is the phenomenon of light emission due to interband transitions when a semiconductor or insulator is heated. This effect was first discovered in quartz. It is different from the familiar black-body glow radiation known as incandescence. Now we will investigate other mechanisms for luminescence in materials. These include photoluminescence – optically stimulated emission of light, and electroluminescence – electrically stimulated emission of light.

9.4.1 Photoluminescence: phosphorescence and fluorescence

How do phosphorescent and fluorescent materials work?

Whenever an electron is excited into a higher energy state it must eventually revert to a lower unoccupied state, and this occurs with the emission of a photon. If the initial excitation is by incident light this process is called photo-luminescence. Phosphorescent materials are used widely for dials on clocks and watches because they glow in the dark. By comparison fluorescent materials glow in the light and are used for 'day glow' colours.

The lifetime of the electrons in the higher energy states determines the duration of the emission process. If the lifetime is short then the emission of photons occurs almost immediately and the luminescence stops when the light source is switched off. This process is called fluorescence. If the lifetime extends over a period of several seconds, a few minutes or even hours then the lumine-scence continues even after the light source is removed. This process is called phosphorescence.

Both of these processes are caused by spontaneous emission of light. That is to say there is no underlying mechanism to stimulate the reversion of the electrons to lower energy states other than spontaneous transition. In general, the wavelengths of the light emitted in fluorescence and phosphorescence are

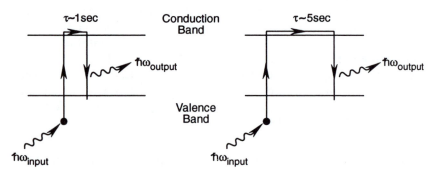

Fig. 9.13 Fluorescence and phosporescence: electron transition diagrams. Large τ corresponds to phosphorescence, small τ to fluoresence.

different from the wavelengths of the incident light, and are usually of a well-defined wavelength which is determined by the band gap energy.

9.4.2 Electroluminescence

How can light emission be stimulated by an electric field?

Optical emission, or luminescence, in solids can be caused by a variety of mechanisms. These are classified by the method of excitation. We have studied the two best known mechanisms, thermal stimulation, (thermoluminescence), and light stimulation, (photoluminescence). A third method is electroluminescence which is the excitation of electrons by an electric field [10]. This is used in the creation of semiconductor light sources.

We will consider injection electroluminescence in a single p-n junction of a semiconductor. The electronic properties of such a junction have been discussed in section 7.8. When the n-type and p-type materials are placed in contact, electrons flow into the p-type material leaving it with a negative charge and the n-type with a positive charge.

If the p-type side of such a junction is then connected to the positive terminal of a voltage supply, current is carried by the flow of electrons into the p-type material where there are already free holes in equilibrium. Recombination of electrons and holes can take place and this results in emission of photons.

In a material such as GaAs, which is a direct gap semiconductor, the absorption edge rises very rapidly with photon energy so that the probability of radiative recombination is very high. In a positive p-sided p-n junction with low impurity concentrations strong emission of light occurs at low temperatures. At higher impurity concentrations, however, the conductivity of the material is too high and emission occurs principally by conduction band–impurity site

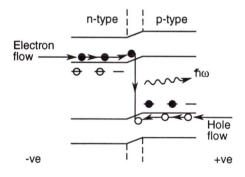

Forward bias

Fig. 9.14 Electroluminescence: electron transition diagram in a forward-biased p-n junction. The electrons are injected from the n-type material.

transitions. These result in the emission of photons of much lower energy, usually in the infra-red region of the spectrum.

The peak of the emission increases to higher photon energies as the current through the junction increases until at a critical current the emission peak sharpens considerably and laser action can begin. We will discuss the operation of these solid-state lasers in Chapter 12.

REFERENCES

1. Wooten, F. (1972) *Optical Properties of Solids*, Academic Press.
2. Abeles, F. (1972) *Optical Properties of Solids*, North Holland, Amsterdam.
3. Hummel, R. E. (1971) *Optische Eigenschaften von Metallen und Legierungen*, Springer, Berlin.
4. Abeles, F. (1966) Optical properties and electronic structure of metals and alloys, in *Proc. Int. Conf.*, North Holland, Amsterdam.
5. Ehrenreich, H. *et al.* (1965) *IEEE Spectrum*, **2**, 162.
6. Brust, D. *et al.* (1964) *Phys. Rev. A*, **134**, 1337.
7. Jiles, D. C. and Staines, M. P. (1983) *Solid State Comms.*, **47**, 37.
8. Swanson, J. G. and Montgomery, V. (1990) *J. Electr. Mater.*, **19**, 13.
9. DiMarco, M. and Swanson, J. G. (1992) *J. Electr. Mater.*, **21**, 619.
10. Wu, M. C. and Chen, C. W. (1992) *J. Electr. Mater.*, **21**, 977.

FURTHER READING

Svelto O. (1982) *Principles of Lasers*, 2nd edn, Plenum, London and New York.

EXERCISES

Exercise 9.1　Optical properties of metals and insulators. The optical constants n and k of four different materials are given in Table 9.2. From these values determine for each material the attenuation coefficient α, the penetration depth δ, the normal reflectance R, the polarization ε_1, and the absorption ε_2. Determine whether each of these materials is a metal or an insulator based on these optical properties.

Table 9.2　Optical constants at $\lambda = 1240$ nm.

Material	n	k (at $\lambda = 1240\,nm$)
1	1.21	12.46
2	0.13	8.03
3	1.51	1.12×10^{-6}
4	1.92	1.5×10^{-6}

Exercise 9.2 Classification of principal electronic transitions. Discuss the principal electronic transitions which can occur in solids and relate them to the band structure diagram. Explain the characteristic colours of materials both in reflectance and transmission in terms of the electronic structure.

Exercise 9.3 Identification of material from optical absorption spectrum. The optical spectrum of an unknown material is given in Fig. 9.15. State whether, on the basis of this data, the material is a metal or an insulator. Determine the absorption threshold for the material and deduce for which optical wavelengths the material is transparent and for which wavelengths it is opaque.

Using the data in Table 9.3 below determine the identity of the material assuming it is one of the three shown.

Table 9.3

Property	A	B	C
Electron mobility	0.15	0.39	0.85
Electrical conductivity (at 300 K) $\Omega^{-1} m^{-1}$	9×10^{-4}	2.2	1×10^{-6}
Effective mass of electrons at 4.2 K (m^*/m)	0.98	1.64	0.07
Band gap (eV)	1.1	0.7	1.4
Absorption edge (nm)	1104	1873	871

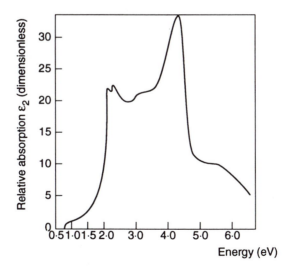

Fig. 9.15 Optical spectrum of unknown material.

10

Magnetic Properties of Materials

In this chapter we look at the magnetic properties of materials. The magnetic properties are a special subgroup of the electronic properties of materials which really form a separate subject. Nevertheless they can also be considered as an integral part of the electronic properties of materials. The most important and interesting magnetic state of a material is known as ferromagnetism. In this case the relative permeability can be very high. This makes these materials useful in transformers and inductors. Another property of ferromagnets is their retention of magnetization. This is utilized in permanent magnets for both motors and generators. In addition particulate magnetic materials are used for magnetic recording purposes. This application happens to form a very large market, both for magnetic materials and the associated electronic support systems for magnetic recording.

10.1 MAGNETISM IN MATERIALS

What causes magnetism in some materials?

The magnetic properties of materials arise almost exclusively from the motion of the electrons. This motion, in the form of electron spin and electron orbital motion, generates a magnetic moment associated with the electron. Much weaker magnetic moments arise from the nucleus, but these are three orders of magnitude smaller. Compare for example the size of the nuclear magneton $\mu_n = 5.051 \times 10^{-27}$ A m^2 with the Bohr (electron) magneton $\mu_B = 9.274 \times 10^{-24}$ A m^2.

There are two theories of the origin of magnetization or bulk magnetic moment in solids which represent limiting or extreme cases. These are the localized or atomic theory, and the itinerant or band theory. In the localized model, the electronic magnetic moments are considered to be bound to the ionic cores in the solid. Such a model applies to the lanthanide series of elements in which the 'magnetic' electrons are inner 4f electrons which are closely bound to the nuclei.

In the itinerant model, the magnetic moments are considered to be due to conduction band electrons which originate as the outer electrons on the isolated atoms. When the atoms are brought together, as in a solid, these electrons move freely throughout the material. This model is considered by some authors to be more appropriate for the 3d transition elements iron, cobalt, and nickel. In reality even in the 3d series metals the itinerant electrons spend more time close to the nuclei, and so the actual situation is somewhere between these extreme or limiting models.

Before proceeding further with these ideas, however, we will need a few definitions.

10.1.1   Magnetic field and magnetic induction

How is a magnetic field generated?

A magnetic field is generated whenever there is electric charge in motion. We denote this field with the symbol $\boldsymbol{H}$. The magnetic field generated by an elemental length of conductor $\boldsymbol{dl}$ carrying a current I is given by the Biot–Savart law,

$$\mathrm{d}\boldsymbol{H} = \frac{1}{4\pi r^2} I\,\mathrm{d}\boldsymbol{l} \times \boldsymbol{u}$$

where r is the radial distance from the conductor at which $\mathrm{d}\boldsymbol{H}$ is measured, $\boldsymbol{u}$ is a unit vector along the radial direction and $\mathrm{d}\boldsymbol{H}$ is the elemental contribution to the total field at r.

The magnetic induction, denoted $\boldsymbol{B}$, is the response of a medium to the presence of a magnetic field. Therefore, for a given field strength $\boldsymbol{H}$, the magnetic induction can be different in different media. The relationship between the magnetic induction and the magnetic field is called the permeability μ of the medium

$$\boldsymbol{B} = \mu\boldsymbol{H}.$$

We should note immediately that μ is not necessarily constant for a material, although in most cases it is either constant or nearly so. The important exception is the class of ferromagnetic materials for which μ varies over an extremely wide range.

The permeability of free space is determined, on the basis of our choice of the metre, newton and ampere as units, to be $4\pi \times 10^{-7}$ henry per metre ($\mathrm{V\,s\,A^{-1}\,m^{-1}}$) and is denoted by the symbol μ_0. Therefore in free space,

$$\boldsymbol{B} = \mu_0\boldsymbol{H}.$$

10.1.2   Magnetization

How do we measure the magnetic response of a material?

When the individual magnetic moments associated with the electrons in a solid are collectively aligned, perhaps by the action of an external magnetic field $\boldsymbol{H}$, we speak of magnetization. We define the magnetization as the magnetic moment per unit volume and denote it by the symbol M. The magnetization increases as more electronic magnetic moments are aligned in the same direction. When all magnetic moments within a solid are aligned in the same direction, the magnetization cannot get any higher. We therefore call this the saturation magnetization.

The magnetization M contributes together with the magnetic field H, to the magnetic induction B. Therefore we can write the totally general equation relating M, H and B,

$$B = \mu_0(H + M)$$

where $\mu_0 H$ is the induction which would be generated by the field H in free space, and $\mu_0 M$ is the additional induction contributed by the presence of the magnetic material. The magnetization is measured in units of amps per metre. Some authors discuss the magnetization in terms of a bound surface current. We should clearly state that this 'bound current' is a fictitious current which is merely invoked as a convenience so that M and H can be treated equivalently.

We defined the permeability μ in section 1.6.3 as the ratio of magnetic induction to field

$$\mu = \frac{B}{H}.$$

Similarly we defined the magnetic susceptibility χ section 1.6.4 as the ratio of magnetization to field

$$\chi = \frac{M}{H}$$

and in the SI system of units $\mu = \mu_0(1 + \chi)$.

Table 10.1   Susceptibilities and relative permeabilities

Material	χ	μ/μ_0
Bi	-1.7×10^{-4}	0.99983
Be	-1.9×10^{-5}	0.99998
Ag	-2.0×10^{-5}	0.99998
Au	-2.7×10^{-5}	0.99997
Ge	-0.6×10^{-5}	0.99999
Cu	-0.8×10^{-5}	0.99999
β-Sn	0.2×10^{-6}	1.00000
W	6.0×10^{-5}	1.00006
Al	2.0×10^{-5}	1.00002
Pt	2.1×10^{-4}	1.00021
Mn	8.3×10^{-4}	1.00083
Fe	$\sim 5 \times 10^3$	$\sim 5 \times 10^3$

10.1.3 Typical values of permeability and susceptibility

What values of permeability and susceptibility do various materials have?

Table 10.1 gives the susceptibilities and relative permeabilities of some dia-magnets, paramagnets and ferromagnets.

10.2 TYPES OF MAGNETIC MATERIAL

How are magnetic materials classified?

There are several different types of magnetic materials but we shall break them down into three traditional categories:

1. diamagnets $\chi < 0$; $\mu_r < 1$
2. paramagnets $\chi > 0$; $\mu_r \geq 1$
3. ordered magnetic materials (e.g. ferromagnets for which $\chi \gg 0$; $\mu_r \gg 1$).

This categorization is of course rather an oversimplification of the different types of magnetic ordering but is still used in traditional magnetism texts. The ordered magnetic materials consist of several subcategories which include: ferro-magnets, ferrimagnets, superparamagnets and even two subcategories with low permeabilities, the antiferromagnets and helimagnets.

10.2.1 Diamagnets

How do diamagnets respond to a magnetic field?

These are materials which have no permanent net magnetic moment on their atoms. In other words, the electrons are all paired with spins antiparallel. When a magnetic field H is applied, the orbits of the electrons change in accordance with Lenz's law, and they set up an orbital magnetic moment which opposes the field. Since this moment is in the opposite direction to the field in diamagnets the susceptibility is negative

$$\chi < 0.$$

The classical theory of diamagnetism was worked out by Langevin [1] and has been discussed by Cullity [2] and Chen [3].

10.2.2 Paramagnets

How do paramagnets respond to a magnetic field?

Paramagnets are materials which have a net magnetic moment per atom due to an unpaired electron spin. In zero field these individual magnetic moments are randomly aligned, but under the action of an external field H they can be aligned in the field direction. As a result of this alignment of moments in the field direction the magnetization M is parallel to the field and hence the susceptibility is positive

$$\chi > 0.$$

In general, however, very large fields are needed to align all the moments and so the susceptibility, although positive, is usually very small, having a typical value of $\chi \approx 10^{-5}$.

10.2.3   Ordered magnetic materials

What other types of magnetic materials are there?

The third class of magnetic materials are the most interesting. These are the ordered magnetic materials the most important of which are the ferromagnets. These include iron, cobalt, and nickel and their alloys and compounds, and several of the rare earth elements, notably gadolinium, and their alloys and compounds. Other ordered magnetic materials include the antiferromagnets, chromium and manganese, the ferrimagnets such as iron oxide, and the helimagnets such as dysprosium, terbium, holmium and erbium.

10.2.4   Curie and Neel temperatures

What happens when the temperature of a magnetic material is raised?

The ordered state of any ferromagnet or ferrimagnet breaks down at a temperature known as the Curie point T_c. Above this temperature the material is disordered, that is the electronic magnetic moments point in random directions even on a local scale. Values of T_c for some materials are shown in Table 10.2.

In antiferromagnets and helimagnets the ordering temperature is known as the Neel point T_N. Values of the Neel temperature for some materials are given in Table 10.3.

Table 10.2 Curie temperatures of various ferromagnets

Material	T_c
Iron	770 °C
Cobalt	1131 °C
Nickel	358 °C
Gadolinium	20 °C

Table 10.3 Neel temperatures of various helimagnets and antiferromagnets

Material	T_N
Dysprosium	$-93\,°C$
Terbium	$-43\,°C$
Chromium	$35\,°C$
Manganese	$-173\,°C$

10.3 MICROSCOPIC CLASSIFICATION OF MAGNETIC MATERIALS

How are the electronic magnetic moments arranged in the various different magnetic materials?

The macroscopic classification into the three traditional groups, based on the permeability values, needs significant modification when we consider the magnetic ordering on the sub-microscopic scale of a few atoms. The microscopic classification needs to include the types of order shown in Fig. 10.1

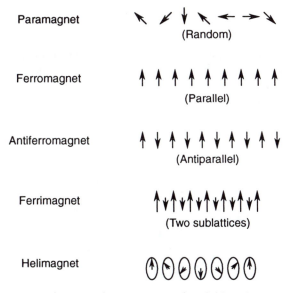

Fig. 10.1 Arrangement of magnetic moments of neighbouring atoms of a one-dimensional lattice for various types of magnetic order.

10.3.1   Electron magnetic moments

Where are the magnetic moments located in a material?

If we consider the classical picture of an electron orbiting a nucleus and also spinning on its axis, as shown in Fig. 10.2, we have charge in motion. Consequently we must have a contribution to the magnetic field and magnetic induction arising through the Biot–Savart law.

In general terms this is correct, however the classical picture has many flaws in it. We will soon find that the numerical values of spin magnetic moment, which is the most significant contribution, differ from the expected value.

In reality there is no electric current here in the classical sense. Therefore the classical model of the electronic magnetic moment merely serves to remind us that there is some link between the angular momentum of an electron and its magnetic moment. However, since the angular momentum of the electron is a quantum phenomenon, it is hardly surprising that the classical prediction breaks down.

We are left with an empirical relation between angular momentum p and magnetic moment m,

$$m = \gamma p$$

where the coefficient of proportionality γ is the gyromagnetic ratio. An alternative form of this relation, in terms of the Bohr magneton μ_B and the Lande splitting factor g is,

$$m = \frac{-g\mu_B}{\hbar} p$$

where $g = 2$ for electron spin alone, and $g = 1$ for electron orbital motion alone. In reality the value of g lies between 1 and 2 in all cases, indicating some orbital and some spin contributions. In the 3d series elements iron, cobalt and nickel the orbital contribution is negligible.

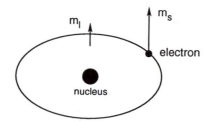

Fig. 10.2 Classical model of electron orbiting a nucleus with orbital magnetic moment m_l and spin magnetic moment m_s.

10.3.2   Order–disorder transitions and the Curie point

What factors determine the Curie temperature of a material?

The ordered magnetic structure which exists in a ferromagnet can be destroyed by raising the temperature. The thermal energy added to the material has a tendency to randomize the orientation of the magnetic moments, while the internal exchange interaction tries to keep them aligned. Eventually a sufficiently high temperature is reached when the thermal energy overcomes the exchange energy and the material undergoes an order–disorder transition.

All ordered magnetic materials (including ferromagnets, ferrimagnets, antiferromagnets, and helimagnets) can be made paramagnetic at a sufficiently high temperature. However not all paramagnets can be converted to ordered magnetic materials by cooling.

The temperature at which the order–disorder transition occurs in ferromagnets is known by convention as the Curie temperature T_c. (However the existence of such a transition temperature was known well before Curie's time, and was even mentioned by Gilbert in his treatise, De Magnete, in 1600!). In antiferromagnets and helimagnets, the order–disorder transition temperature is known as the Neel temperature. The behaviour of the susceptibility χ changes at these ordering temperatures and we say that it exhibits 'critical behaviour' close to T_c or T_N.

Some materials have both a Curie and a Neel temperature because they exhibit more than one ordered magnetic phase. Examples are terbium ($T_N = 230\,\text{K}$, $T_c = 220\,\text{K}$) and dysprosium ($T_N = 180\,\text{K}$, $T_c = 85\,\text{K}$), which undergo transitions paramagnetic $\rightarrow$ helimagnetic $\rightarrow$ ferromagnetic as the temperature is reduced.

10.3.3   Temperature dependence of susceptibility

How does the susceptibility of a magnetic material change with temperature, and how can this be quantitatively described?

The Curie and Curie–Weiss laws were empirical discoveries of the temperature dependence of the paramagnetic susceptibility of certain magnetic materials. It is worth noting that these laws are not as widely applicable as is often generally supposed. However, their simple form and their subsequent explanation using classical statistical thermodynamics means that they hold an important place in the historical development of our understanding of magnetism.

The Curie law [4] states that the susceptibility χ of a paramagnet is proportional to the reciprocal of the temperature T in degrees Kelvin,

$$\chi = \frac{C}{T}$$

where C is a constant.

The Curie–Weiss law [5] is a generalization of the Curie law to include those materials which undergo an order–disorder transition to ferromagnetism or ferrimagnetism at T_c. In these cases the susceptibility in the paramagnetic phase is also inversely proportional to the temperature according to the relation,

$$\chi = \frac{C}{T - T_c}$$

where C is a constant.

It is important to note that the Curie–Weiss law only applies to the susceptibility of a magnetic material in its paramagnetic phase.

10.3.4   The Curie law

Can the Curie law be explained in terms of the statistical behaviour of an array of individual magnetic moments?

The Curie law can be explained on the local moment model using classical Maxwell–Boltzmann statistics. In materials with unpaired electrons, there is a net (or resultant) magnetic moment per atom m which is the vector sum of the spin and orbital magnetic moments. The energy of this moment in a magnetic field H is

$$E = -\mu_0 \boldsymbol{m} \cdot \boldsymbol{H}.$$

If we suppose that the magnetic moments are non-interacting and use classical statistics, then the probability of an electron occupying an energy state E is

$$P(E) = P_0 \exp(-E/k_B T).$$

If there are N magnetic moments per unit volume, then the magnetization M, which is the overall magnetic moment per unit volume, will be found by integration,

$$M = \int_0^N \boldsymbol{m} \cos \theta \, dn$$

$$M = N\boldsymbol{m} \frac{\int_0^\pi \cos \theta \sin \theta \exp\left(\frac{\mu_0 mH \cos \theta}{k_B T}\right) d\theta}{\int_0^\pi \sin \theta \exp\left(\frac{\mu_0 mH \cos \theta}{k_B T}\right) d\theta}$$

$$M = N\boldsymbol{m} \left\{ \coth\left(\frac{\mu_0 mH}{k_B T}\right) - \frac{k_B T}{\mu_0 mH} \right\}$$

and saturation magnetization occurs when all moments are aligned parallel. We use here the symbol M_0 to denote saturation magnetization, as distinct from spontaneous magnetization or technical saturation in a ferro- or ferrimagnet, which we shall denote M_s (section 10.5.4). Since

$$M_0 = Nm$$

then

$$M = M_0 \left\{ \coth\left(\frac{\mu_0 mH}{k_B T}\right) - \frac{k_B T}{\mu_0 mH} \right\}.$$

This expression is the Langevin equation for classical paramagnetism based on the local moment model. We can derive the Curie law directly from this. The Langevin function can be expressed as an infinite series in $\mu_0 mH/k_B T$. For high temperatures we find that $\mu_0 mH/k_B T \ll 1$, so that the first term in the series dominates

$$M = Nm \left\{ \frac{\mu_0 mH}{3k_B T} \right\}$$

and substituting $C = N\mu_0 m^2/3k_B$, this gives,

$$M = \frac{CH}{T}$$

and, since $\chi = M/H$, we obtain

$$\chi = \frac{C}{T}$$

which is the Curie law.

10.4 BAND ELECTRON THEORY OF MAGNETISM

Are the 'magnetic' electrons localized on the ionic sites or are they free to move throughout the material?

The itinerant theory of magnetism attributes the magnetic properties of materials to unpaired electrons in the conduction band. These electrons by definition can migrate throughout the whole material and are therefore 'itinerant'. This interpretation is valid for some materials but not for others. It does seem to be broadly applicable to the magnetism of the 3d series elements such as iron, cobalt and nickel.

10.4.1 Pauli paramagnetism

How can we derive an expression for the paramagnetic susceptibility in terms of the behaviour of conduction electrons?

The itinerant conduction band theory of paramagnetism was developed by Pauli [6]. This theory leads to a temperature-independent paramagnetic sus-

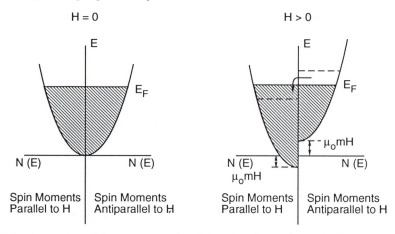

Fig. 10.3 Occupation of electron energy levels in a free-electron material in zero magnetic field and under an applied magnetic field **H**.

ceptibility. Beginning from our earlier discussion of electron bands, consider for example the parabolic free-electron band shown in Fig. 10.3. For ease of visualization, we represent the spin-up states on the left and the spin-down states on the right. These we will term the spin-up and spin-down half-bands. Electrons will occupy the lowest available energy states. In the absence of a field, the energy levels of the spin-up and spin-down states are degenerate. Therefore, the two half-bands are symmetric, occupancies of the two half-bands are equal, and the net magnetic moment per atom is zero.

When a field is applied the individual electronic moments acquire an additional energy ΔE which depends on the scalar product of the magnetic moment m with the field H

$$\Delta E = - \mu_0 m \cdot H.$$

This energy lifts the degeneracy of the two half-bands, because the spin-up bands move to lower energy while the spin-down bands move to higher energy. As a result, the Fermi energy equalizes in the two half-bands and some electrons switch from spin down to spin up. This leads to a net magnetic moment in the spin-up direction. It is also clear from this diagram that only those electrons that were close to the Fermi level will be able to switch direction.

10.4.2 Dependence of magnetization on field in Pauli paramagnetism

How does the susceptibility depend on the Fermi level, number of conduction electrons and magnetic moment per electron?

From our earlier consideration of electron states at the Fermi level, only a fraction T/T_F of the conduction electrons can contribute to the magnetization. This again is similar to the fraction of electrons that contribute to the heat capacity and electrical conductivity. Therefore according to the classical Curie law we should expect,

$$M = \frac{N^* \mu_0 m^2 H}{3 k_B T}$$

where M is the magnetization and N^* is the number of electrons per unit volume that can change the orientation of their spins. Since $N^* = NT/T_F$ (as described in section 4.5.1) where N is the total number of conduction electrons per unit volume this leads to,

$$M = \frac{N \mu_0 m^2 H}{3 k_B T} \left(\frac{T}{T_F} \right)$$

where m is the electronic magnetic moment, and $k_B T$ is the Boltzmann energy.

If we approach the problem in a more exact way, using quantum mechanics and the band theory of electrons, the above result is modified slightly, but ultimately we arrive at the same form of expression for M. The number density of electrons parallel to the field N_+ is,

$$N_+ = \int f(E) D(E + \mu_0 m H) dE$$

$$N_+ \approx \int f(E) D(E) dE + \mu_0 m H \, D(E_F)$$

and the number density antiparallel to the field N_- is, by a similar argument

$$N_- \approx \int f(E) D(E) dE - \mu_0 m H \, D(E_F).$$

The magnetization is therefore given by,

$$M = m(N_+ - N_-)$$
$$= 2 m^2 \mu_0 H D(E_F)$$

where $D(E_F)$ is the density of states at the Fermi level. We can see that the resulting magnetization M is dependent not only on the applied field H, but also on the density of states at the Fermi level $D(E_F)$. Using $2D(E_F) = 3N/2k_B T_F$ [7 p. 415],

$$M = \frac{3N}{2 k_B T_F} m^2 \mu_0 H$$

and therefore the Pauli free-electron band theory of paramagnetism leads to the following equation for the susceptibility,

$$\chi = \frac{M}{H}$$

$$= \frac{3Nm^2\mu_0}{2k_B T_F}.$$

This gives a temperature-independent paramagnetic susceptibility which is observed in a number of metals such as sodium, potassium and rubidium.

10.4.3 Electron band model of ferromagnetism

How can we generalize the above theory of magnetic susceptibility to describe ferromagnetic materials?

The band theory of ferromagnetism is a simple extension of Pauli's band theory of paramagnetism to ferromagnets with the inclusion of an exchange interaction (internal effective magnetic field) to align the electrons in a co-operative manner in the absence of an external applied field. This causes a relative displacement of the spin-up and spin-down half-bands known as the exchange splitting. It is qualitatively similar to that encountered under the action of an applied magnetic field in Pauli paramagnetism, except that here the shift in energy is much larger and occurs in the absence of an applied magnetic field. The net spontaneous magnetization of a material is again determined by the difference in occupancy between the spin-up and spin-down states.

In the 3d series elements, the outer electron bands of interest, which contribute to the magnetic properties, are the 3d (total capacity 10 electrons) and the 4s (total capacity 2 electrons). In iron, cobalt, and nickel, the 4s band is completely filled with its complement of two electrons. Therefore, since this level must be occupied with a spin-up and a spin-down electron it can have no contribution to the magnetic moment per unit volume. The magnetic properties are therefore determined by the partially filled 3d band.

Suppose that we have a material such as nickel consisting of atoms with 8 electrons each in a partially filled 3d band which has a capacity of 10 electrons. Then if we consider 10^{29} atoms per m^3 of this material, each atom could have up to 10 of these 3d electrons in the band, and so there will be one electron band consisting of 5×10^{29} energy levels with spin-up and an equal number of energy levels with spin-down states (giving a total of 10^{30} possible electron states). There will be 8×10^{29} electrons in this 3d band and these will occupy the lowest energy states available. With no interactions between the electrons, they will be equally distributed between the two sub-bands ($3d^\uparrow$ and $3d^\downarrow$) with no net magnetic moment. So in this case we will again have a Pauli paramagnet.

In order to get the net imbalance of spins necessary for ferromagnetism it is essential to invoke an exchange energy which displaces the energies of the spin-up and spin-down half-bands even in the absence of an external field.

10.4.4   Exchange coupling

Why should there be a net magnetic moment in a partially filled electron band of a ferromagnet?

In a partially filled energy band, it is possible to have an imbalance of spins. This is caused by the presence of an exchange interaction between the electrons in the conduction band which has the effect of aligning the spins parallel. This exchange interaction is quantum mechanical in origin and has no classical analogue. However, it is sometimes modelled as a classical mean field [8]. When this effective field is calculated it is found to be extremely large, being equivalent to a magnetic field of about $10^9 \, A \, m^{-1}$ (10^7 Oe).

The exchange interaction has the effect of reducing the energy associated with parallel alignment of spins even in the absence of an external field. Therefore the occupancy of the spin-up state becomes energetically favoured over the spin-down state in zero field. This results in a net magnetic moment.

10.4.5   Spin-up and spin-down half-bands

Under what conditions can the exchange interaction lead to a net magnetic moment?

The exchange energy can only alter the alignment of the moments if the reduction in energy due to exchange is greater than the energy difference between the lowest available spin-up state and the highest occupied spin-down state. In other words the system of spins will always configure itself to the lowest possible energy state, taking the exchange energy into account. If the exchange energy is present but is not large enough to alter the ground state in this way then no net magnetic moment will arise (Fig. 10.4).

The energy resulting from the exchange interaction, E_{ex}, is usually represented by the Heisenberg model [9] as,

$$E_{ex} = -J_{ex}s_1s_2$$

where J_{ex} is the exchange operator, and s_1 and s_2 are the spins on electrons. When $J_{ex} > 0$, then we have a tendency to parallel alignment which minimizes the exchange energy and leads to ferromagnetism, but this can only occur if

$$|E_{ex}| \geqslant \Delta E$$

where ΔE is the energy difference between the lowest available spin-up state and the highest occupied spin-down state. This condition simply ensures that

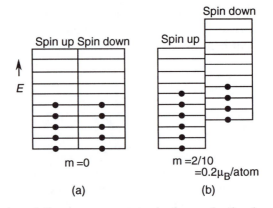

Fig. 10.4 Occupation of discrete energy states in the conduction band of an itinerant-electron ferromagnet. In (a) no exchange splitting gives equal occupancy of the spin-up and spin-down half-bands, while in (b) the exchange splitting leads to an imbalance of spins and a net magnetic moment per atom.

any change in orientation of electronic magnetic moment causes a reduction in the total energy of the system.

10.4.6 Magnetic moment per atom

How is it possible to obtain magnetic moments per atom which are non-integral multiples of the Bohr magneton?

In the example of nickel given above, suppose that in unit volume 3×10^{28} out of the 4×10^{29} (i.e. 3 out of 40) electrons from the spin-down half-band transfer to the spin-up half-band because it is energetically favourable. This results in a net excess of 6×10^{28} spins, or a net magnetic moment of 0.6 Bohr magnetons per atom. We see from this how it becomes possible to explain fractional numbers of Bohr magnetons per atom by the band theory of ferromagnetism.

The electrons fill the spin-up band first as shown in Fig. 10.5. When the half-bands overlap electrons can be added to the spin-down half-band before the spin-up half-band is filled. A complete separation of the energies of the spin-up and spin-down bands would leave the entire spin-up half-band at a lower energy than the spin-down half-band, and therefore all the electrons would fill the spin-up half-band first.

If there were more states in the spin-up band than there were electrons available to fill them, there would be no electrons in the spin-down band. Since in this example each atom contributes one electron to the spin-up band, the net result would be a magnetic moment of one Bohr magneton per atom. The actual electron band structure for nickel is shown in Fig. 10.6 in which the exchange-split states are shown as pairs of 'parallel' half-bands. In this case the two half-bands overlap.

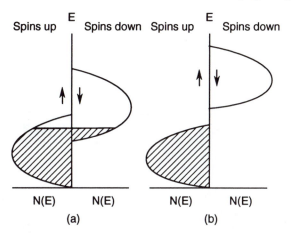

Fig. 10.5 Occupation of electron energy levels in a ferromagnet. The two half-bands are split by the exchange coupling. In (a) the exchange splitting does not completely separate the half-bands, while in (b) the exchange splitting is large enough to cause complete separation.

10.4.7 Magnetic moments in Fe, Co and Ni

What are the typical distributions of the outer electrons in iron, cobalt and nickel?

Since it is the outer, unpaired, electrons which contribute to the magnetic properties of the 3d series elements we will consider here the 3d and 4s outer electrons only. The magnetic moments per atom of these materials when in an aggregate (i.e. solid form) are different from the isolated atoms. This is shown in Table 10.4. The isolated atoms of course do not have 'continuous' electron bands, instead they have discrete energy levels.

10.4.8 Rigid-band model for transition metal alloys

What effect does alloying have on the magnetic moment per atom of the transition metal alloys?

When considering the magnetic properties of the 3d transition metals and alloys, the electronic structure of the conduction band can be approximated by

Table 10.4 Distribution of 3d and 4s electrons in iron, cobalt and nickel

Material	Number of 3d electrons	Number of 4s electrons	Net atomic moment (Bohr magnetons)
Fe	6	2	2.2
Co	7	2	1.7
Ni	8	2	0.6

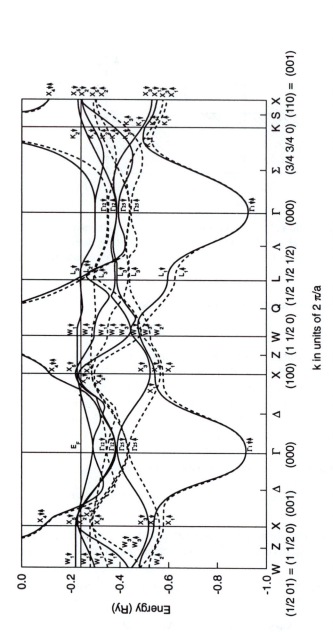

Fig. 10.6 Electron band structure of ferromagnetic nickel after Wang and Callaway [10]. Reproduced with permission from C. S. Wang and J. H. Callaway, *Phys. Rev.*, **B9**, 1974, p. 4897.

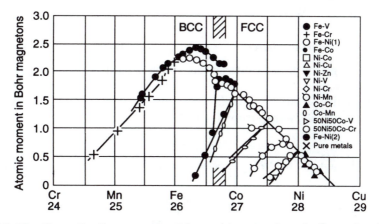

Fig. 10.7 The Slater–Pauling curve for 3d transition metals and alloys, showing the variation of net atomic magnetic moment as a function of alloy composition.

assuming that the electron bands are rigid even on addition of other 3d-alloying elements. Therefore by alloying with 3d elements which have more, or fewer, electrons than the principal component of the alloy, the effect is simply to add, or subtract, electrons from the existing band. Addition or subtraction of electrons simply depends on whether the added element lies to the left or right of the principal component element in the periodic table.

In the 3d transition metals and alloys this occurs without substantially altering the energy levels of the bands. The net result is the variation of magnetic moment per atom with alloy composition according to an empirical model known as the Slater–Pauling curve [11, 12] (Fig. 10.7). This describes the magnetic moment in terms of the number of unpaired conduction electrons per atom.

On increasing the number of electrons per atom from zero the magnetic moment at first increases, as the spin-up band is filled, and then decreases, once the spin-down band starts to fill faster than the spin-up band. The assumption of the rigid band approximation which underlies the Slater–Pauling curve should not be taken too literally, but it does seem to work as a first approximation.

10.5 THE LOCALIZED ELECTRON MODEL OF MAGNETISM

Is there an alternative to the itinerant electron theory of magnetism?

In many cases the electronic magnetic moments can be considered to be localized at the atomic or ionic sites. This is particularly appropriate for the lanthanide series of elements in which the 4f 'magnetic' electrons are closely bound to the atomic cores.

10.5.1   The Curie–Weiss law

How can we explain the difference between the Curie–Weiss law and the Curie law?

By introducing a mean field interaction, which is used merely to represent the quantum mechanical exchange interaction in a simplified form, the Curie–Weiss law can be derived from the Curie law. Suppose the interaction can be expressed as a field H_e, which is proportional to M

$$H_e = \alpha M.$$

The total magnetic field H_{Tot} experienced by an individual magnetic moment is then

$$H_{Tot} = H + H_e$$
$$= H + \alpha M.$$

The variation of magnetization M with total field H_{Tot} should still obey Curie's law,

$$\frac{M}{H_{Tot}} = \frac{M}{H + \alpha M}$$

but now the measured susceptibility is,

$$\chi = \frac{M}{H}$$

$$= \frac{C}{T - \alpha C}$$

$$= \frac{C}{T - T_c}$$

where $T_c = \alpha N \mu_0 m^2 / 3 k_B$ is the Curie temperature. This means that the Curie temperature is directly related to the strength of the exchange interaction as measured by the mean field parameter α.

10.5.2   Classical theory of ferromagnetism

How does the exchange interaction field explain the existence of ferromagnetism?

The mean field interaction was introduced by Weiss [5, 8] to account for the observed paramagnetic susceptibility of magnetic materials which undergo a transition to ferromagnetism at the Curie point. This idea can be carried over into the ferromagnetic regime where it explains the existence of magnetic order.

A brief outline of the classical Weiss theory follows. Suppose that any magnetic moment m_i experiences an interaction with any other moment m_j in the material.

This can then be expressed as an effective magnetic field experienced by the ith moment,

$$H_{e_{ij}} = \alpha_{ij} m_j / V_j$$

where α_{ij} is the interaction between the ith and jth moments and V_j is the volume occupied by m_j.

The total internal interaction field is then given by the sum of the individual interaction fields over the whole material. Assuming that $V_j = V$ for all moments,

$$H_e = \frac{1}{V} \sum_j \alpha_{ij} m_j.$$

This makes it energetically favourable for the electron magnetic moments to align parallel when $\alpha_{ij} > 0$, leading to ferromagnetism.

10.5.3 The mean field approximation

What form does the exchange field take if all interactions are assumed equal?

If we assume that the interactions between all moments are identical (i.e. if we make the mean field approximation) with $\alpha_{ij} = \alpha$ for all pairs i, j, then the above summation is simplified to,

$$H_e = \frac{1}{V} \sum_j \alpha m_j$$

$$= \frac{\alpha}{V} \sum_j m_j$$

and if M_s is the spontaneous magnetization throughout the volume under consideration (usually a single domain), then,

$$H_e = \alpha (M_s - m_i / V)$$

and since $m_i / V \ll M_s$, we can replace $M_s - m_i / V$ with M_s,

$$H_e \simeq \alpha M_s$$

which is the form of interaction envisaged in a ferromagnetic material in the original theory by Weiss.

10.5.4 Magnetic order, spontaneous magnetization and domains

If the magnetic moments are all aligned parallel in ferromagnetism, then how can a ferromagnet ever be demagnetized?

We have discussed the fundamental difference between ferromagnets and paramagnets which is the existence of long-range order in ferromagnets. This means that large numbers of magnetic moments are aligned parallel. This is true of iron, cobalt, and nickel at room temperature.

This leads to an apparent contradiction. Most specimens of iron, cobalt, and nickel do not have a bulk magnetization of the size expected on this basis unless they have been deliberately 'magnetized', that is exposed to a strong external magnetic field. The contradiction is resolved through the existence of magnetic domains which are microscopic volumes (usually quite small, being typically $0.001 \, mm^3$ in volume) in which all magnetic moments are aligned parallel. The direction of this magnetization is different from domain to domain, leading to a low value of bulk magnetization.

We therefore need to distinguish between the macroscopic (or bulk) magnetization which occurs after a ferromagnet has been subjected to a magnetic field, and the microscopic (or spontaneous) magnetization which is the magnetization within a domain. We shall denote the bulk magnetization M and the saturation magnetization M_0. The spontaneous magnetization M_s is close in value to M_0 and as the temperature T is reduced to absolute zero, $M_s(T)$ approaches M_0.

Whereas M can be affected by the presence of an external field, M_s is largely unaffected and M_0 is completely unaffected. The change in magnetization M caused by an external magnetic field arises from reorientation of domain magnetizations, causing the individual domain magnetic moments to line up throughout the material.

10.6 APPLICATIONS OF MAGNETIC MATERIALS

What uses do magnetic materials find?

Magnetic materials find their main applications in three areas. Soft magnetic materials [13, 14], which have high permeability and low coercivity, are used in electromagnets, electric motors and inductor cores in which the primary objective is to generate as much magnetic induction B as possible under the action of a magnetic field H. Hard magnetic materials [15], which have low permeability but high coercivity and remanence, are used in permanent magnet applications in which the materials are used to generate a magnetic induction without a conventional electrical power supply and also in permanent magnet motors. Finally, magnetic particles and magnetic thin films are used in magnetic recording of information in the form of analogue signals or digital data. All of these applications have been discussed in detail in the literature [16].

REFERENCES

1. Langevin, P. (1905) *Ann. Chem. et Phys.*, **5**, 70.
2. Cullity, B. D. (1972) *Introduction to Magnetic Materials*, Addison Wesley, Reading, Mass.
3. Chen, C. W. (1977) *Magnetism and Metallurgy of Soft Magnetic Materials*, North Holland, Amsterdam.
4. Curie, P. (1895) *Ann. Chem. Phys.*, **5**, 289.
5. Weiss, P. (1906) *Compte Rend.*, **143**, 1136.

6. Pauli, W. (1926) *Z. Physik*, **41**, 81.
7. Kittel, C. (1986) *Introduction to Solid State Physics*, 6th edn, John Wiley & Sons, New York.
8. Weiss, P. (1907) *J. de Phys.*, **6**, 661.
9. Heisenberg, W. (1928) *Z. Physik*, **49**, 619.
10. Wang, C. S. and Callaway, J. (1974) *Phys. Rev. B.*, **9**, 4897.
11. Slater, J. C. (1937) *J. Appl. Phys.*, **8**, 385.
12. Pauling, L. (1938) *Phys. Rev.*, **54**, 899.
13. Boll, R. (1979) *Soft Magnetic Materials*, Heyden & Son, London.
14. Snelling, E. C. (1988) *Soft Ferrites*, 2nd edn, Butterworths, London.
15. Parker, R. J. (1990) *Advances in Permanent Magnetism*, John Wiley & Sons, New York.
16. Jiles, D. C. (1991) *Introduction to Magnetism and Magnetic Materials*, Chapman & Hall, London.

FURTHER READING

Chikazumi, S. (1964) *Physics of Magnetism*, John Wiley & Son, New York.
Jiles D. C. (1991) *Introduction to Magnetism and Magnetic Materials*, Chapman & Hall, London.

EXERCISES

Exercise 10.1   Strength of the exchange field in iron. Iron has a Curie temperature of 1043 K and a magnetic moment of 2.2 Bohr magnetons per ion. Find the strength of its internal exchange field.

Exercise 10.2   Comparison of the magnetic moments on atoms in bulk form and in isolation. Compare the known saturation magnetizations of iron, cobalt and nickel in bulk with the known magnetic moment of the free atoms or ions and comment on the result. Calculate the occupancy of the 3d and 4s bands in iron from these results.

Exercise 10.3   Spontaneous magnetization and the exchange field. Assuming that the Heisenberg exchange interaction can be introduced simply as an effective field that is proportional to the spontaneous magnetization within a domain, derive an expression for the magnetization of a domain as a function of magnetic field H, magnetic moment per atom m and temperature T starting from the classical Langevin equation for paramagnetism.

   If $m = 2 \times 10^{-23}$ A m^2, $N = 9 \times 10^{28}$ m^{-3} and $T = 300$ K, find the value of the mean field parameter α which is needed in order to cause spontaneous magnetization.

Part Three

Applications of Electronic Materials

11

Microelectronics – Semiconductor Technology

The next few chapters provide overviews of some selected applications of electronic materials. In this chapter we look at materials for microelectronics applications. Microelectronics itself is a vitally important and wide-ranging group of related fields that includes semiconductor materials technology. In a book such as this it is only possible to skim the surface of such a diverse subject. Therefore the limited objective here is to discuss some of the considerations which go into selecting materials for microelectronics applications. This is based on the specific electronic properties of the materials which have been discussed in previous chapters. We show here how the relative advantages of silicon, germanium and gallium arsenide need to be considered for specific applications. Our main focus remains on the material properties rather than the functional requirements.

11.1 USE OF MATERIALS FOR SPECIFIC ELECTRONIC FUNCTIONS

What is 'microelectronics'?

When we come to the subject of microelectronics, we are concerned mainly with semiconductors [1, 2] and semiconductor devices [3]. This is an extremely diverse area of technology which extends from solid-state physics (understanding the electronic properties), through materials science (fabrication of micro-electronic circuits), to electrical engineering (performance of components and devices) and computer engineering (integration of large numbers of devices and their application in digital systems). The entire field of microelectronics can be traced back to one source, the development of the transistor at Bell Telephone Laboratories in 1947 by Bardeen, Brattain and Shockley [4]. By the mid-1980s the electronics industry had become the largest single manufacturing industry in the US.

Semiconductors are used to produce the key components in the majority of electronic systems including communications, data processing, control and

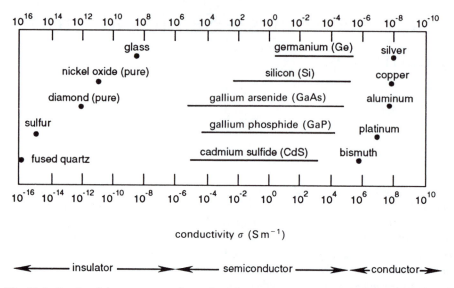

Fig. 11.1 Conductivity ranges of semiconductors compared with insulators and conductors.

consumer electronics equipment. Of the two main classes of semiconducting materials, intrinsic and extrinsic, it is the extrinsic semiconductors which are of greater technological importance here. The reason for this is that the electrical conductivity can be carefully controlled in extrinsic semiconductors using dopants, whereas in intrinsic semiconductors such fine control is difficult if not impossible to achieve.

11.1.1   Control of electronic properties of semiconductors

What is a dopant?

A dopant is an impurity element which is deliberately added to the semiconductor to change its conductivity. Dopants can provide extra electrons (e.g. P or As in silicon) to form n-type semiconductors, or they can provide extra holes (e.g. Ga or Al in silicon) which are positive charge carriers to form p-type semiconductors. Doping concentrations are typically a few parts in a million and this makes it possible to fabricate devices with desired electrical properties. The range of possible conductivities of semiconductors is shown in Fig. 11.1.

11.2 SEMICONDUCTOR MATERIALS

Which are the main semiconductor materials currently available?

The first semiconductor material in widespread use was germanium but this exhibits relatively high leakage currents at moderately elevated temperatures, and it was superseded by silicon in the 1960s. Silicon has replaced germanium for almost all applications because of its lower leakage current, resulting from its wider band gap, and the fact that high-quality silicon dioxide can be easily produced to form a good insulating layer on the material where necessary. The silicon dioxide layer strongly adheres to the surface of the silicon and can be used as a mask. In addition silicon is more abundant than germanium and therefore cheaper.

The electrical properties of silicon are also fairly easy to control. The variation of resistivity of silicon with impurity concentration in shown in Fig. 11.2.

The primary material of interest in this chapter is therefore silicon, since most of the semiconductor industry is built around this material and its oxides. It constitutes 95% of all semiconductor hardware sold world-wide at present. For use in electronic devices 'electronics grade' silicon is produced. This is in the form of single crystals of silicon obtained by slowly withdrawing seed crystals from molten silicon.

11.2.1 Alloy semiconductors

What makes gallium arsenide and related materials so special?

Gallium arsenide [5] and other direct band gap semiconductors are also of interest. Together with other related III–V compound semiconductors it is likely to have enormous impact on the optoelectronics and computer industries in the future because of the possibility of fabricating fast electronic devices from

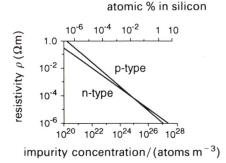

Fig. 11.2 Variation of the resistivity of silicon with impurity concentration. Reproduced with permission from N. Braithwaite and G. Weaver, *Electronic Materials*, published by Butterworths, 1990.

these materials, and because of its ability to support optical functions. The III–V compounds are used in laser diodes, solar cells and light-emitting diodes which are discussed in the next chapter. Gallium arsenide and the other III–V semiconductors are the subject of intensive research [6] because of this high potential. Continuing interest stems from the following unusual combination of properties:

1. high band gap energy which can be engineered (i.e. altered) by combining it with other materials (e.g. InP) to allow optical transitions over a range of energies;
2. direct band gap, making it suitable for optoelectronics applications;
3. high electron mobility leading to very fast operation of GaAs devices.

11.2.2   The III–V semiconducting compounds

How can the band gap of these semiconductors be controlled?

One interesting property of these compounds is that they can be mixed together to form solid solutions. For example in gallium arsenide, aluminium or indium can be substituted for some or all of the gallium; or phosphorus or antimony can be substituted for some or all of the arsenic. This can be used to make subtle changes in the electrical and optical properties of the material. All III–V compounds have the same crystal structure and similar values of lattice constant which is advantageous for fine control of the electronic properties. Composition of the III–V compounds can be carefully controlled by modern materials processing techniques (e.g. growth by molecular-beam epitaxy) which allows the band gap to be tailored for particular applications. This can result for example in materials which emit light over a range of wavelengths close to the optical range which can be used for light-emitting diodes of various colours, or for semiconductor lasers.

These materials provide an important avenue of investigation for the development of new optoelectronic devices. The main difficulty is that the III–V compounds are much more difficult to process than silicon and the raw materials are more expensive too. This leads to a higher cost of devices. However, the possibility of using them for optical applications, and the higher electron mobilities, which lead to fast electronic devices, can outweigh the extra cost. So for specialized applications III–V compound semiconductors are finding a market.

11.3   TYPICAL SEMICONDUCTOR DEVICES

What kinds of devices are fabricated on wafers of semiconductor, and how do these work?

We begin by developing an understanding of the electronic processes in some very simple cases. Then we will go on to discuss applications which can be presented without further detailed reference to band structures. The simple devices that we shall consider in order to develop this understanding are the p-n junction and the transistor. Our first objective is merely to describe how these work, using as a basis, our understanding of the behaviour of electrons in materials which has been developed in earlier chapters. Only then will we be concerned with their possible applications.

11.3.1 The p-n junction

What does the electron band structure look like in the vicinity of a semiconductor junction between p- and n-type material?

Junctions between semiconducting materials are crucial for applications of these materials because in this way very diverse electronic properties can be produced. The simplest of these is the p-n junction, which we have discussed previously in section 7.8. The electronic behaviour of the p-n junction, which consists of two semiconductors with different conduction mechanisms and different Fermi levels joined together, can lead to some quite interesting results. These provide a simple basis for understanding more complex devices. The schematic electron band structure is shown in Fig. 11.3.

As a result of differences in the Fermi levels on each side of the junction, electrons flow from the n-type material to the p-type material. Consequently the n-type material becomes positively charged and the p-type becomes negatively charged. Eventually the Fermi levels become equalized in the two materials. The region at the interface, which becomes depleted of mobile charge carriers as the mobile electrons flow into the p-type material and the mobile holes flow into the n-type material, is known as the depletion layer, also sometimes called the space charge region. This is shown in Fig. 11.4. The p-n junction can be forward or reverse biased by connecting a power supply to the two pieces of semiconductor, as shown in Fig. 11.5.

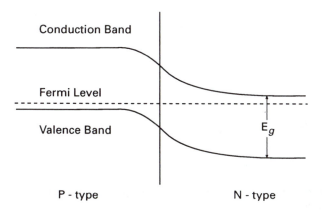

Fig. 11.3 Band structure diagram for a p-n junction.

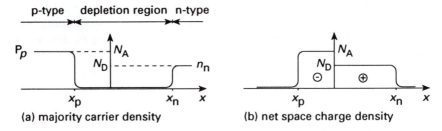

(a) majority carrier density   (b) net space charge density

Fig. 11.4 Electron and hole densities on either side of a p-n junction.

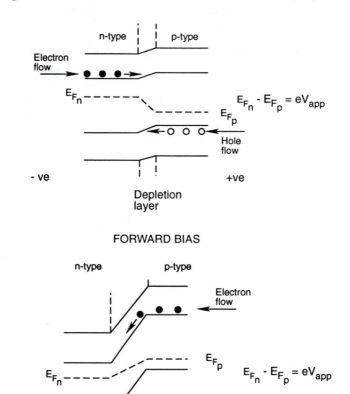

FORWARD BIAS

REVERSE BIAS

Fig. 11.5 Connection of a p-n junction to a DC power supply which results in biasing of the junction.

11.3.2 Performance characteristics of a p-n junction

When subjected to a voltage how does the current through the p-n junction change?

Any electrons trying to move from n-type to p-type encounter a potential barrier as shown in Fig. 11.5, while those electrons trying to move from p-type to n-type pass easily down the potential ramp. The current/voltage characteristics of a p-n junction are shown in Fig. 11.6. It can be seen that the junction acts like a diode, in which the current rises steeply with applied voltage in one direction (forward bias) but the device remains non-conducting in the other direction (reverse bias).

It should be noted however, that the current/voltage characteristics of a p-n junction are strongly temperature dependent. By raising the temperature from 20 °C to 50 °C the reverse-bias current can be raised by more than an order of magnitude, as a result of thermal stimulation of more electrons across the band gap, giving a higher charge carrier density in the conduction band.

11.3.3 Transistors

How does a transistor work?

A transistor is a solid-state device for amplifying, controlling and generating electrical signals (voltages and currents). The whole field of microelectronics really began with the discovery of the transistor [4]. This device was the proto-type of many important devices and we shall pause briefly to discuss its operation. The transistor consists of two semiconductor junctions, with one semiconductor region common to both (Fig. 11.7). There are two possible configurations, 'n-p-n'

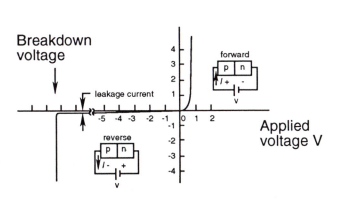

Fig. 11.6 Current/voltage characteristics of a p-n junction.

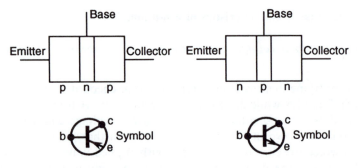

Fig. 11.7 Configurations for a transistor. Both involve sandwiching one type of semiconductor between pieces of another tupe of semiconductor.

and 'p-n-p'. The first transistors were made of germanium because at the time it could be produced in pure form more easily than other semiconducting materials, such as silicon.

The region in the middle of this semiconducting 'sandwich' is the *base*. The other two are known as the *emitter* and *collector*. All three have direct electrical connections in a circuit.

11.3.4   Band structures of junction transistors

What are the special characteristics of the electron band energy levels in the vicinity of the transistor junctions?

In a p-n-p transistor the valence and conduction bands of the centre or n-type material are lower than for the p-type material on the outside. This is shown in Fig. 11.8.

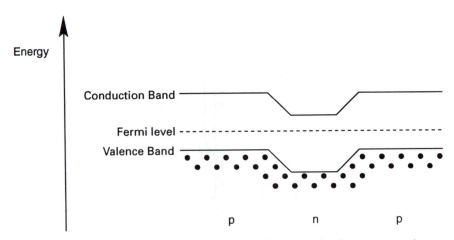

Fig. 11.8 Relative energies of conduction and valence bands of a p-n-p transistor.

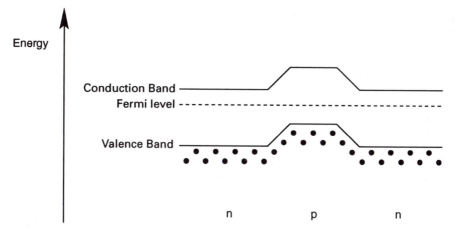

Fig. 11.9 Relative energies of conduction and valence bands of an n-p-n transistor.

In an n-p-n transistor the situation is just the opposite. In this case the energy levels of the centre p-type region are higher than those of the outer n-type region as shown in Fig. 11.9.

11.3.5 Effects of biasing the n-p-n transistor

What happens to the electron band structure of an n-p-n transistor when subjected to a bias voltage?

As can be seen from the case of the p-n junction biasing the junction can radically affect the current voltage characteristics. The same is true for a transistor. The

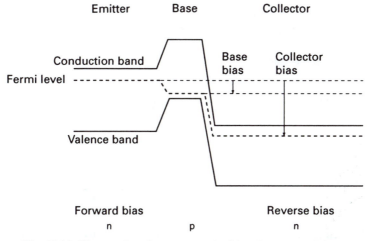

Fig. 11.10 Electron band structure of a biased n-p-n transistor.

electron band structure of a transistor with the emitter–base junction forward biased and the collector–base junction reverse biased is shown in Fig. 11.10.

11.3.6   Typical transistor characteristics

How can a transistor be controlled by the bias voltage so that it acts as an amplifier?

A transistor can be used to amplify small signal voltages or currents. We will consider an n-p-n transistor as an example. The principles of operation remain the same for a p-n-p transistor. The emitter–base is forward biased while the base–collector is reverse biased. Electrons climb the barrier between emitter and base being pushed up the potential by the applied voltage, as shown in Fig. 11.11. They then rush down the potential ramp from base to collector, as shown in Fig. 11.12, producing a high current.

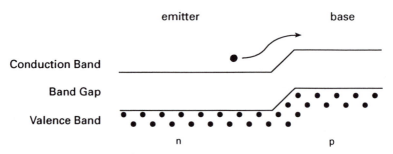

Fig. 11.11 Electrons being pushed up the potential ramp from the emitter to the base of an n-p-n transistor.

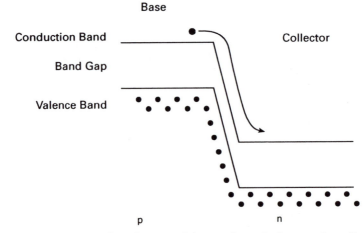

Fig. 11.12 Electrons descending the potential ramp from the base to the collector in an n-p-n transistor.

Of the three principal configurations of a transistor [2, p. 135] we will consider only the 'common-base' and 'common-emitter' arrangements. In the common base configuration the input voltage is V_{EB} between the emitter and base, and the output voltage is V_{CB} between the collector and base. The current gain is then simply the change in collector current for a given change in emitter current

$$\alpha = \frac{\Delta I_C}{\Delta I_E}$$

Since the collector and emitter currents are virtually the same in these devices, as explained below, this means that the gain α is close to unity.

In the common emitter configuration the input voltage V_{BE} is between the base and emitter and the output voltage V_{CE} is between the collector and emitter. In this case the current gain is the change in collector current for a given change in base current

$$\beta = \frac{\Delta I_C}{\Delta I_B}$$

This gain is usually much larger than α. By conservation of charge it follows that,

$$I_B + I_C + I_E = 0$$

and therefore,

$$\Delta I_B = -(\Delta I_C + \Delta I_E)$$

and hence

$$\Delta I_B = -\Delta I_C \frac{(\alpha - 1)}{\alpha}$$

leading to

$$\beta = \frac{\Delta I_C}{\Delta I_B} = \frac{\alpha}{(1 - \alpha)}$$

which gives a relationship between the common base and common emitter gains. Since $\alpha \simeq 1$ this means that β is very large by comparison, being typically in the range 10–1000. Therefore, a small change in the base current can lead to a large change in the collector current.

The explanation of the behaviour of the device is as follows. An increase in the base voltage reduces the potential barrier between the emitter and base regions allowing more electrons to pass into the base. The base region itself is very thin in these devices so that a very high proportion of the electrons entering from the emitter pass in to the collector without recombining with holes in the base. This leads to a collector current which is comparable to the emitter current. A large increase in collector current can therefore result from a small increase in base current.

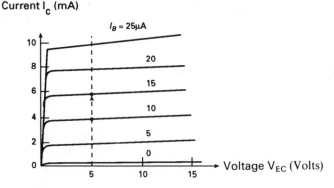

Fig. 11.13 Current/voltage characteristics for a p-n-p transistor in the common emitter configuration.

The base current is typically of the order of microamps, while the emitter and collector currents are typically of the order of milliamps. The voltage/current characteristics of an n-p-n transistor in the common emitter configuration are shown in Fig. 11.13.

11.4   MICROELECTRONIC SEMICONDUCTOR DEVICES

How many such simple devices can be fabricated on a single silicon chip?

We have seen that it is possible to produce devices with interesting and diverse properties by altering the electronic structure of materials. Two simple examples were the p-n junction and the transistor, but these are only the simplest of devices that can be fabricated from semiconductor materials such as silicon. Large numbers of these devices can be produced and interconnected on a single wafer of silicon [7, 8].

This can result in literally millions of such simple electronic devices being fabricated on a single silicon chip of about 4 mm square. Recent developments by Intel announced at the end of 1992 have increased the 1.2 million transistors on the commercial 486DX microprocessor to 3.2 million on the 78 mm² surface of its latest 'Pentium' microprocessor.

11.4.1   Integrated circuits

What is an 'integrated' circuit?

An integrated circuit (IC) is an assembly of interconnected electrical components which is fabricated as a single unit on a substrate of semiconducting material which is usually silicon. The IC consists of an assembly of circuit elements, both passive (capacitors and resistors) and active (transistors and diodes). These are fabricated together with the necessary electrical connections. These circuit elements are arranged in such a way that the whole integrated circuit performs an electrical circuit function. The integrated circuit was invented by Kilby at Texas Instruments in 1958 and independently by Hoerni and Noyce of Fairchild Semiconductor Corporation at about the same time.

The first integrated circuits built in 1958 contained 10 components fabricated on a single chip of semiconductor. By 1970 the number of components on the same size chip had risen to 1000. Modern ICs are typically 5 mm square and large numbers of identical ICs are fabricated simultaneously on silicon wafers 100 mm in diameter which are subsequently separated into individual chips.

11.4.2 Microprocessors

What is a microprocessor?

Large-scale integration was developed during the 1970s and this enabled thousands of transistors to be packed onto a silicon chip 3 mm square. This gave rise to the microprocessor, an advanced integrated circuit which, in addition to the conventional circuit functions, contains the arithmetic, logic and control circuitry necessary to carry out the functions of a central processing unit (CPU) of a digital computer. The microprocessor was invented by Intel Corporation in 1971.

Very large-scale integration (VLSI) was developed in the 1980s. This has further increased the density of devices on the chip. By the early 1990s microprocessors were fabricated with more than 3 million transistors on a single silicon chip less than 5 mm square (Fig. 11.14).

11.4.3 Fabrication

How are the devices actually made out of a single block of silicon?

The fabrication process for producing devices on silicon wafers is quite complicated and involves a large number of steps. The complete process takes several days to complete. However we will simplify the process into its principal stages, taking as an example the production of a metal oxide semiconductor field effect transistor (MOSFET). This proceeds as outlined below.

1. Production of single-crystal silicon wafer.
2. Doping of material, either by diffusion of impurities or by ion-implantation, followed by diffusion.
3. Growth of epitaxial layers on surface of wafer, such as lightly doped silicon on a heavily doped substrate. This is the stage at which the material is supplied by the wafer manufacturer.

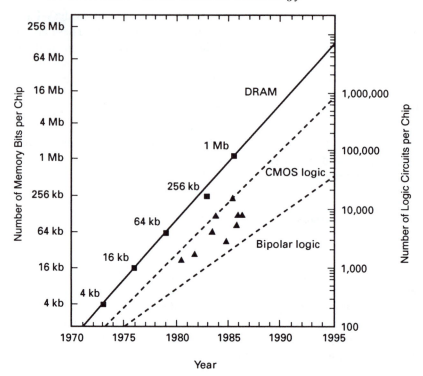

Fig. 11.14 Increase in the number of memory bits per chip from 1970 to 1995 (projected).

4. Ion-implantation into the surface to create regions with different electrical properties such as n-type and p-type regions.
5. Subsequent heat treatments to allow implanted dopants to diffuse further into the material to establish the desired well structure.
6. A thin 0.03 μm layer (gate oxide) of silicon dioxide is produced over the surface, followed by a thin 0.1 μm nitride layer.
7. Photolithography is used to define which areas on the surface are to be etched and which are to be left with the nitride layer. The nitride areas are for sources, drains and channels (electrically active areas).
8. A thicker oxide layer 0.6 μm (field oxide) is then deposited on areas without the nitride coating, since these areas are to be covered by insulator.
9. The nitride layer is removed by the use of orthophosphoric acid leaving only the thin 'gate oxide' layer in the chosen regions.
10. The silicon–silicon dioxide interface contains fixed positive charges in the oxide. These are compensated by negative charges just inside the silicon. Since these positive charges are desirable, ion-implantation of acceptor atoms (for example BF_2^+) in the silicon is used to ensure that it is easier to get mobile positive charges into the oxide layer [9].

11. Polycrystalline silicon gates are aligned at the active sites. The silicon is heavily doped to increase its conductivity. These gates are deposited over the thin 'gate oxide' regions. This stage is known as gate deposition.

12. Photoresist layers are then placed over the surface in two steps. First the surface, except for the p-wells with their heavily doped polysilicon gates, is covered. Then arsenic atoms are ion-implanted deep enough to penetrate the areas covered only by the thin oxide layers in the p-wells. Next the surface, except for the n-wells, also with their heavily doped polysilicon gates, is covered again with photoresist. Then boron atoms are ion-implanted into the areas covered only by the thin oxide layer in the n-wells.

13. The photoresist layer is removed using sulphuric acid and hydrogen peroxide. This leaves an uneven surface with a top layer of about 1 μm of silicon dioxide with some contact holes in the surface for electrical connections.

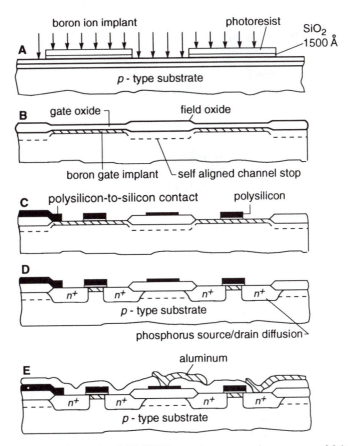

Fig. 11.15 Section through a MOSFET transistor at various stages of fabrication.

14. The planarization stage involves the use of an organic silicate in alcohol solution which is cured at 800 °C to leave a fairly smooth glassy surface ready for metallization.

15. Once the surface has been smoothed, metal layers can adhere to it and these layers are laid down to form the electrical connections. This process is called metallization. The deposition involves sputtering of a metallic surface layer using conventional lithography to obtain the required pattern. The rate of deposition is typically 0.01 μm per second. The metal layer usually consists of a 0.15 μm titanium–tungsten layer to prevent aluminum migrating into the silicon, and on top of this 0.5 μm layer of Al–4%Cu.

Some of the principal stages in the fabrication process are shown in Fig. 11.15. In the future direct wafer bonding [10] will be used for silicon-on-insulator materials, bipolar devices and for small semiconductor sensors.

11.5 FUTURE IMPROVEMENTS IN SEMICONDUCTORS

What are the directions in which semiconductors are likely to develop in the near future?

The present-day microelectronics industry is built on the fabrication of micro-scopic electrical circuits on silicon. These circuits are then sealed, packaged to protect them, and then electrical connections to the outside world are added. A typical microelectronic circuit, the proverbial 'computer chip', appears as shown in Fig. 11.16.

One direction for further development is in the refinement of techniques for fabrication, particularly in view of the sensitivity of the electronic properties of semiconductors to impurities. This refinement includes the elimination of

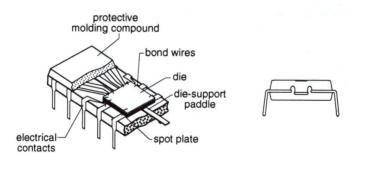

Cut away section   External end view

Fig. 11.16 Cut-away section through an integrated circuit package showing the silicon chip together with the necessary circuit connections. The whole is encapsulated in a protective plastic housing.

impurity elements and the elimination of flaws in the material. Another direction is the use of amorphous silicon [11] instead of the present single-crystal silicon for certain applications. This has distinct advantages in terms of reduced cost. The main application area for amorphous silicon is in low-cost, high-efficiency photovoltaic cells.

The need for newer, faster semiconductors seems to be continuing apace. Single-element semiconductors based on the group IV elements germanium and silicon are now developed to full maturity, which means there is little scope for further radical improvement using these materials alone. Compound semiconductors composed of elements from groups III and V of the periodic table have been under investigation for some years [12]. Gallium arsenide comes into this category. This material can be used to fabricate high-speed devices because of its high electron mobility (low effective mass of electrons). The detection and generation of light by the semiconductor is another great advantage leading to its use in optoelectronic applications for which there is an increasing demand as will be discussed in Chapter 12.

11.5.1 Gallium arsenide technology

What are the relative advantages and disadvantages of gallium arsenide over silicon?

Gallium arsenide circuits are faster than similar silicon circuits, they consume less power, have lower noise, can radiate and detect light more efficiently in the visible wavelength range and have an easily engineered band gap. For example when alloyed with gallium phosphide the band gap can be altered from 1.42 to 2.26 eV corresponding to wavelengths in the optical range from red to green. In view of all these advantages it might well be asked why there has not been a wholesale switch from silicon technology to gallium arsenide technology in the semiconductor industry. The reasons however are clear: gallium arsenide is both expensive and difficult to fabricate into integrated, single-chip circuits and devices, while silicon is very plentiful and is relatively easily fabricated.

On the other hand the most rapidly developing applications of semiconductors are in photonic transmission of information ('optical computing') and this has led to a high level of interest in gallium arsenide. Modern computers can produce data streams which are too fast for conventional copper wire conductors to transmit over distances greater than about 200 metres. The result of conventional transmission under these conditions is that consecutive signals begin to blur. Optical communication overcomes this limitation and allows computers to communicate at high speed over very large distances using optical fibres and light-emitting diodes or diode lasers. It is in this particular application that the optical semiconductors are finding important uses, since despite the difficulty of fabrication they provide functions that other semiconductors cannot.

11.5.2   Band gap engineering

Since the band gap is the dominant characteristic of a semiconductor how can materials with predetermined band gaps be fabricated?

A large band gap in a semiconductor eliminates much of the intrinsic contribution to the conductivity, which was discussed in section 7.2, by making thermal stimulation of electrons across the band gap less probable. The larger the band gap the lower the number density of electrons in the conduction band. This means that the conducting properties of the semiconductor can be more precisely controlled by the addition of donor or acceptor elements if the band gap is larger.

The III–V semiconductor compounds include GaAlIn–AsPSb which can be alloyed together to form solid solutions. This allows the band gap in these materials to be varied simply by changing the chemical composition [13]. The band gaps of the various alloys are shown in Fig. 11.17.

One of the first alloy semiconductors of this group to be used was the ternary compound $GaAs_{1-x}P_x$ which was used for red light-emitting diodes. The maximum brightness of this alloy as an LED was found to occur with x = 0.4 which gave a band gap of 1.9 eV. Other compound III–V semiconductors used for optical applications include $Al_xGa_{1-x}As$ and $Ga_xIn_{1-x}As_yP_{1-y}$. The ability to change the band gap, as in the GaAs–GaP and GaAs–AlAs compounds

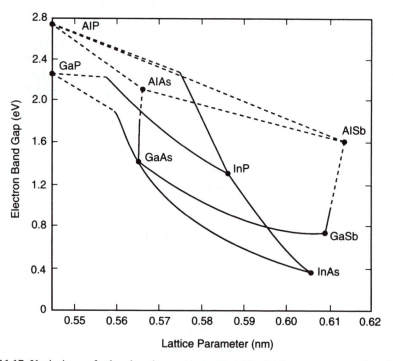

Fig. 11.17 Variation of the band gap energy with lattice parameter for III–V semiconductors.

is important for optical communication and emission and detection of light at different wavelengths [14].

11.5.3 Speed of operation

How important is the speed of response of a semiconductor device?

Another issue is the speed of semiconductor devices. Improved response of devices in terms of the time taken for the device to respond to an input can be a critical issue in some applications. In the civilian sector of the economy there is a demand for ever faster computers and communications. In the case of gallium arsenide the high speed arises from the high curvature of the electron bands near the Γ point which results in a low effective mass of the electrons and hence a high electron mobility as discussed in sections 5.4.5 and 7.7. Electron mobilities also depend on the temperature and impurity content, generally decreasing as both of these increase.

REFERENCES

1. Smith, R. A. (1978) *Semiconductors*, 2nd edn, Cambridge University Press.
2. Sze, S. M. (1981) *Physics of Semiconductors Devices*, 2nd edn, John Wiley and Sons, New York.
3. Allison, J. (1990) *Electronic Engineering Semiconductors and Devices*, McGraw-Hill, London.
4. Bardeen, J. and Brattain, W. H. (1948) *Phys. Rev.*, **74**, 230; (1949) **75**, 1208.
5. Peaker, A. R. (ed.) (1990) *Properties of Gallium Arsenide*, IEE Inspec, London.
6. Brodsky, M. H. (1990) *Sci. Am.*, **262** (2), 68.
7. Ghandhi, S. K. (1983) VLSI fabrication principles, John Wiley and Sons, New York.
8. McDonald, J. F. (1984) *IEEE Spectrum*, **21**, 32.
9. Irby, J. *et al.* (1992) *J. Elect. Mater.*, **21**, 543.
10. Bengtsson, S. (1992) *J. Electr. Mater.*, **21**, 841.
11. Madan, A. (1986) *IEEE Spectrum*, **23** (9), 38.
12. Bell, T. E. (1985) *IEEE Spectrum*, **22** (10), 46.
13. Drummond, T. J. (1988) *IEEE Spectrum*, **25** (6), 33.
14. Goodhue, W. D. *et al.* (1990) *J. Electr. Mater.*, **19**, 463.

FURTHER READING

Bell, T. E. (1986) The quest for ballistic action *IEEE Spectrum*, **23** (2), 36.
Holton, W. C. and Cavin, R. K. (1986) A perspective on CMOS technology trends. *Proc. IEEE*, **74**, 1646.
Jones, M. E., Holton, W. C. and Stratton, R. (1982) Semiconductors: the key to computational plenty. *Proc. IEEE*, **70**, 1380.
Madan, A. and Shaw, M. (1987) *Physics of amorphous semiconducting devices*, Academic Press, New York.
Skromme, B. J., Bose, S. S. and Stillman, G. E. (1986) New shallow acceptor levels in GaAs, *J. Electr. Mater.*, **15**, 345.
Williams, C. K. (1992) Kinetics of trapping, detrapping and trap generation, *J. Electr. Mater.*, **21**, 711.

12

Optoelectronics – Solid-State Optical Devices

The purpose of this chapter is to describe one of the fastest developing areas of electronic materials. The need for optical communications between computers, or for telecommunications, is driven by the need for faster and more precise transmission of data. Conventional electrical communications suffer from problems which are related to the speed of communication and the distance over which such communication needs to be made. Optical communications through fibres can meet these requirements. There are four main components to optoelectronics for communications: light generation, transmission, detection and user-interfacing. The devices for these functions are: light-emitting diodes, optical fibres, photodetectors and liquid-crystal displays. This chapter presents a selected overview of some of the materials used for optoelectronics in each of these main areas. Again the treatment aims for breadth rather than depth, so the interested reader should consult more specialized works for details.

12.1 ELECTRONIC MATERIALS WITH OPTICAL FUNCTIONS

What do we mean by 'optoelectronics'?

Optoelectronics is the combination of optical and electronic processes in materials. For example, an electronic transition across the band gap of a semiconductor from conduction band to valence band with the emission of a photon is an optoelectronic process.

In optoelectronics, we are principally concerned with the generation of light as a result of electronic processes in materials, efficient transmission of light for communications purposes and detection of incident light. The main topics of interest here are, therefore, lasers and light-emitting diodes, photodetectors, fibre optics and optical displays.

The whole field of optoelectronics is currently receiving great attention because of the possibility of developing high-speed computers capable of communicating at a rate of more than 10 gigabits per second using optical methods. By comparison, standard electronic computers operate at frequencies of typically 50 megabits per second. The improvement in speed is therefore likely to be up to a factor of 200. Other developments include the ability to produce low-power, low-cost optical displays (LEDs and LCDs) using semiconductors, and the ability to fabricate inexpensive, low-power semiconductor lasers.

12.1.1   Photodetectors

How can we detect light and convert it into an electrical voltage?

A photodetector is a semiconductor device that converts light into an electrical voltage [1]. There are two types of photodetector, the photoconductor and the photodiode. The photoconductor is a material in which electrons are stimulated from the valence band to the conduction band by incident light. This leads to an increase in conductivity which can be related to the incident light intensity and wavelength.

A photodiode is a reverse-biased p-n junction in which the stimulation of electrons from the valence band to the conduction band leads to a current under the action of an electric field (potential gradient). This current is proportional to the intensity of incident light, provided the energy of the incident photons is sufficient to excite electrons across the band gap. The solar cell is an example of such a device. It consists of a silicon p-n junction which, when placed in sunlight, generates an electric current that can flow in an external circuit connecting the two semiconductors.

Photodetectors therefore depend on the photoelectric effect which we have discussed in Chapter 7, in which absorption of photons causes electrons to be stimulated to higher energy levels, leading to either an increase in conductivity, or if a potential gradient is present, as in a p-n junction, to the flow of current. The photoelectric effect in semiconductors causes a marked change in conductivity which is quite different from that observed in metals because, whereas the metals already have large numbers of electrons in the conduction band, the semiconductors do not. The relative effect on the conductivity is therefore much greater in semiconductors.

In discussing the photoelectric effect here we draw a distinction between it and photoemission in which the electrons are actually liberated from the material by the incident photons. Photoemission is characteristic of metals which have higher energy conduction electrons available.

12.1.2   The p-n junction as a detector

How can the simple p-n junction be used as a detector?

In order to act as a detector the p-n junction is reverse biased, since for sensitive detection a large fractional change in the current is required. It is fairly easy

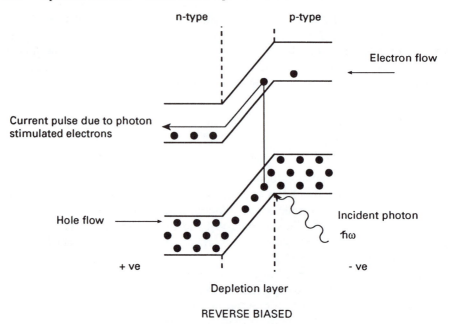

Fig. 12.1 Principle of operation of a photodetector.

to detect a change of $1\,\mu A$ if the current doubles to $2\,\mu A$, but somewhat more difficult to detect a change of $1\,\mu A$ in $1\,mA$. Under reverse-biased conditions relatively little conventional current flows in the p-n junction.

As a result of the reverse bias the energy differential across the junction is enhanced, which makes it very difficult for electrons to pass from the n-type to the p-type material because of the large energy barrier. This ensures that there is a very low dark current, and so any photocurrent appears to be relatively large by comparison. However, electrons in the conduction band of the p-type material can easily pass into the n-type material.

When photons with sufficiently high energy to excite electrons across the band gap ($\hbar\omega > E_{\mathrm{g}}$) are incident on a p-n junction, electrons are stimulated into the conduction band (Fig. 12.1). These are then driven into the n-type material by the potential gradient across the reverse-biased junction. The result is a change in conductivity (increased current for the same applied voltage) and this is used to measure the intensity of the incident light.

12.1.3 Semiconductor light sources

How can a semiconductor be used as a light source?

Conventional light sources emit light as a result of their high temperature in a process known as incandescence. In this case most of the energy supplied results in heat rather than light, so as a light source the process is relatively inefficient. Luminescence is a different process which is the result of electronic excitation in which electrons at a high energy fall to an available state at a lower energy. This results in the emission of a photon of energy equal to the difference in electron energy before and after the event. Light-emitting diodes (LEDs) are used for this purpose.

Semiconductor light sources operate in the inverse way to the semiconductor detectors. Again the device can be a simple p-n junction which operates as a diode. It can act as a light source when forward biased. An LED is a form of luminescent lamp. When current flows through, the semiconductor electrons recombine with holes by decaying to a lower energy level and emitting a photon. LEDs are usually made from III–V compound semiconductors related to gallium arsenide, for example $GaAs_{1-x}P_x$ on GaAs.

An attractive feature of light-emitting diodes is that they perform well, have a long lifetime, do not overheat and in most cases are relatively inexpensive, except for blue LEDs which at present require the rather expensive semiconductor SiC. Currently LEDs do not provide enough light for illumination but are often used for information displays or data presentation. In the future it is quite likely that LEDs will be developed for illumination and in this case, because of their high efficiency and low heat emission, they are likely to become the preferred mode of lighting.

The wavelength of the emitted light is governed by the semiconductor band gap, because the initial high-energy electron state is most often at the bottom of the conduction band and the final low-energy electron state is at the top of the valence band. Optical transitions such as this occur with high probability in direct band gap semiconductors but not in indirect band gap semiconductors. Therefore silicon is not a suitable material for LEDs, but gallium arsenide and the family of III–V compound semiconductors are well suited for this application.

12.2   MATERIALS FOR OPTOELECTRONIC DEVICES

Which materials are widely used for practical optoelectronic devices?

Present-day optoelectronic devices incorporate exotic materials processed with the utmost care, and they function by the manipulation of electrons and holes by electrical and optical means. Both LEDs and laser diodes are used as visual indicators. In most cases the III–V semiconductors, of which GaAs is the archetype [2], are used to make devices which detect [3] or emit [4] light and have rapid response times due to their high electron mobilities.

The III–V semiconductors consist of gallium or aluminium or indium together with arsenic or phosphorus or antimony. Their direct band gap allows optical transitions to occur with high efficiency as phonon assistance is not required to ensure conservation of electron momentum. In addition, these elements can be alloyed to form solid solutions in which the band gap energy is determined by the chemical composition. The variation of band gap energy with composition has been shown in Fig. 11.17.

12.2.1   Band gaps of III–V and II–VI alloy semiconductors

How can the colour of light-emitting diodes be selected?

Most LEDs have emissions in the infra-red (2000–700 nm) or the optical range (700–400 nm). Visible LEDs are used as numeric displays or as small indicator or warning lamps. Infra-red LEDs are used as optical isolators or as sources in optical communications systems such as computer communications.

The usual LED materials are aluminium–gallium arsenide $Al_xGa_{1-x}As$, gallium–indium arsenide–phosphide, $Ga_xIn_{1-x}As_yP_{1-y}$, and gallium arsenide–phosphide $GaAs_yP_{1-y}$. The band gap of these materials is dependent on the alloy composition, and can be engineered to a high precision by alloying. Fortunately the III–V elements form solid solutions which makes alloying simple in these cases and so the band gap is relatively easy to control. This means that the wavelength of the emitted light can be selected by choosing the appropriate alloy composition. Wavelengths throughout the optical range and into the infra-red, are easily available with these alloys. The III–V alloys still have direct band gaps so they remain optically useful.

In gallium arsenide up to 30% of the input energy can be converted to light but this occurs at a wavelength of 900 nm which is in the infra-red, and so is of no use as an optical display – for which the wavelengths of emissions must

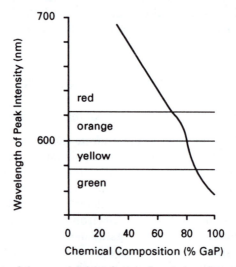

Fig. 12.2 Dependence of the wavelength of optical emissions from GaAs–GaP, the first LED material in widespread use.

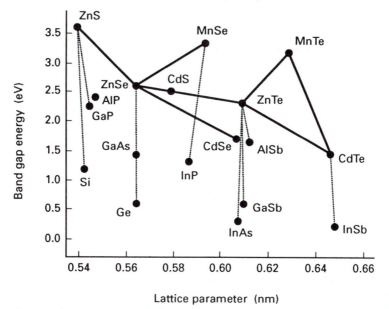

Fig. 12.3 Band gap energies of II–VI compound and alloy semiconductors.

be detectable by the human eye. Gallium phosphide gives visible light at about 500 nm, but its efficiency is low. The alloys of these materials have intermediate wavelengths and efficiencies.

Gallium nitride emits light in the visible range with a blue-green colour, but again with rather low efficiency. The efficiency of $GaAs_{1-x}P_x$ decreases as the composition x increases. Beyond $x = 0.45$ the band gap becomes indirect and the probability of transition, and therefore the optical efficiency, decreases rapidly. However the addition of nitrogen in place of phosphorus leads to a more efficient radiative process.

If emissions at short wavelengths in the blue end of the optical spectrum are required there are two options. Expensive SiC semiconductors can be used. However, the latest development is the II–VI family of semiconducting compounds such as ZnS and ZnSe which have band gaps in the range 3.6 eV–1.5 eV spanning the entire optical spectrum (Fig. 12.3). These materials and others, such as CdS and SrS, are now being investigated for use as full colour, thin-film electro-luminescent displays [5].

12.2.2 Minority carrier injection: injection diodes

How is electron–hole recombination brought about in the light-emitting diode?

These are forward-biased p-n junctions. Electrons are injected from the n-type to the p-type material by the voltage. Once in the p-type material they recombine with holes and emit light. Injection diodes are often used as small indicator lamps. Their great advantage is that they have high efficiency, perform well and in most cases are relatively inexpensive.

The process of minority carrier injection is used in the light-emitting diode. This process occurs at the boundary of a p-n junction if a forward bias voltage is applied. Electrons can be driven into the p-type material, and holes into the n-type material by the electric field caused by the voltage from the power supply. This process is used in the LED because it leads to recombination of electrons and holes with the emission of light.

12.2.3 Recombination process and light emission in LEDs

Where does the electron–hole recombination occur?

The forward biasing of the p-n junction (Fig. 12.4) forces electrons up the potential ramp of the conduction band so that they reach locations in k-space where lower energy states are available in the valence band. An optical transition can then occur from the conduction band into an available state in the valence band. This occurs in the depletion region (space charge region) of the junction. The result is the emission of a photon of energy equal to the band gap. There is also a rapid increase in current with forward-bias voltage.

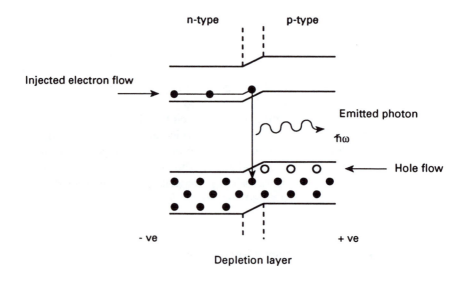

FORWARD BIASED

Fig. 12.4 Principle of operation of a photoemitter.

The wavelength of the emitted light is determined by the band gap of the material. By engineering the band gap of the material, light emitters of different but specific frequencies can be fabricated.

12.2.4 Light detection and generation

What other applications for optoelectronic semiconductors are anticipated?

We have discussed the generation of phase-coherent light-using laser sources. The p-n junction described can also be used to detect and generate noncoherent light. It is suggested that these luminescent semiconductor light sources will be the preferred method of illumination in the future, replacing conventional incandescent filament lamps and fluorescent tubes. They will be more energy efficient, will not generate heat and will have virtually limitless lifetimes. These are significant advantages which are now leading to the development of such light sources for the consumer market.

12.3 LASERS

How does a laser semiconductor light source differ from a normal luminescent semiconductor light source?

A laser is a device which emits an intense beam of light composed of photons all of the same wavelength, and all in phase. This is known as a coherent light source. The laser gives an intense energy concentration because the beam divergence is small as a result of the method of generation of the light in the laser cavity.

There are several different types of lasers:

1. semiconductor lasers
2. optically-pumped solid-state lasers
3. dye lasers
4. gas discharge lasers
5. gas dynamic lasers
6. chemical lasers
7. liquid lasers
8. free-electron lasers.

Of these we will only be concerned here with the semiconductor laser. Semiconductor (junction) lasers are usually aluminium-gallium-arsenide or simply gallium arsenide [6]. When a large electrical current is passed through such a device laser light emerges from the junction.

When an electron has been excited into the conduction band of a semiconductor for example, it must return finally to a lower energy state, either in the valence band or a localized impurity state in the band gap. This occurs with the emission of a light photon. In many cases this emission is spontaneous as in the light-emitting diode. Consequently the radiation is phase incoherent, and therefore this does not constitute laser action. However in a laser this emission of light occurs by controlled stimulation rather than spontaneously, and this gives phase-coherent photon emission which is characteristic of laser light.

12.3.1   Emission of laser light from a semiconductor

How are the conditions for stimulated emission created?

In the very simplest case, consider two energy levels, one at the top of the valence band and the other at the bottom of the conduction band as shown in Fig. 12.5.

Clearly in order to get light emission we must have energy stored somewhere in the system waiting to be released. This is achieved in the form of electrons in excited states. In a laser, the term 'population inversion' is used to describe the presence of a larger number of excited electrons. The lifetime of the electrons in the excited state must also be long enough to ensure that the transition to a lower energy state does not occur spontaneously, but only under controlled conditions. The presence of such a population inversion is a common feature of all lasers, since large numbers of electrons need to be maintained in a high energy state over an extended period of time. This requires rather special conditions which we will discuss.

12.3.2   Production and maintenance of the population inversion

How are large numbers of electrons maintained in a metastable higher energy state?

Optical pumping is used to produce a large number of electrons in a metastable high energy state. This involves a high-intensity light source to stimulate

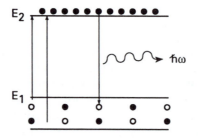

Fig. 12.5 Electron energy levels of a two-level laser.

electrons into the higher energy level. The electrons later decay to their original lower level with the emission of light. When the electron reverts to the lower energy state a photon is emitted with energy equal to the difference in energy of the two states and its frequency is ω given by

$$\hbar\omega = E_2 - E_1.$$

Population inversion is the redistribution of electrons among available energy states by stimulation into metastable higher energy states. This results in more electrons occupying high energy states than can be indefinitely sustained. It represents energy stored in the material. Population inversion can be achieved by a number of mechanisms of which 'optical pumping' is only one. Other methods for producing the population inversion include: electrical discharge, chemical reactions, nuclear reactions, electron beam injection and conventional current injection.

The pumping of electrons into excited states is most effective if there is a broad range of energies in the upper energy band. The pumping can then be achieved by a broad range of wavelengths. This means that the simple two-level laser is not the most efficient configuration for producing laser light from a semiconductor.

If the semiconductor has a range of excited energy levels, then the light source can have a continuous range of wavelengths (i.e. can be polychromatic). Electrons are then excited into a wide range of energy states in the conduction band where they remain until they are caused to de-excite, with the emission of a photon. Typical optical sources for optical pumping are xenon and tungsten iodide filament lamps.

12.3.3 Stimulated emission

How is the stored energy released?

The release of stored energy from the laser occurs when excited electrons drop to a lower energy state as a result of being stimulated by a photon of the correct energy passing through the material. This leads to the emission of a photon with energy $\hbar\omega$ corresponding to the difference in energy of the electron before and after decay.

When electrons occupy an elevated energy state they can be forced to return to a lower unoccupied state by a kind of resonance involving a photon of the same energy as the transition. The emitted light is phase coherent with the stimulating light. In laser action the large number of electrons occupying high-energy metastable states can be caused to return to available lower energy states by an avalanche effect of photons in a chain reaction. This leads to the production of a high-intensity, monochromatic, phase-coherent beam of light. This is a laser beam.

12.3.4   The emission process

Is the emitted photon phase coherent with the stimulating photon?

Experimental evidence shows that the passage of a photon $\hbar\omega$ through the material can stimulate the transition of other electrons to lower energy states with the same transition energy through a resonance. The emitted photon is found experimentally to be phase coherent with the stimulating photon.

If this process continues, further photons are produced with the same energy and in phase coherence with the first photon. The result is light amplification by stimulated emission of radiation (hence the term 'laser'). Laser light is monochromatic because the difference in energy levels, and hence the difference in energy of the electrons before and after emission, are the same throughout the material. Therefore the photons emitted at any location in the material will have the same energy.

Laser light is also strongly collimated as a result of the method of generation in a long narrow cavity (Fig. 12.6). The semiconductor has mirrored end-faces. The laser light is reflected many times between the end mirrors of the cavity leading to a beam with very low divergence. Any photon which is not closely parallel to the axis of the cavity is soon absorbed, leaving only those with low divergence.

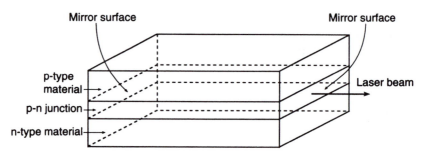

Fig. 12.6 Amplification of light photons by stimulated emission of radiation in the cavity of a laser.

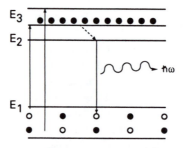

Fig. 12.7 Electron energy band diagram for a three-level laser.

12.3.5   Three-level laser

How can the performance of a laser be improved by the use of semiconductor material with different electronic properties?

The need for high-efficiency pumping and for a well-defined transition energy are mutually exclusive in a two-level system. This is because a well-defined optical transition energy requires the upper electron level to be narrow. The simplest way around this problem is to have three electron levels as shown in Fig. 12.7, with a broad upper energy band, with short lifetime for high pumping efficiency, from which the electrons de-excite into a narrow energy, long-lifetime level. The long lifetime of electrons in this narrow band allows high population inversion to be created and maintained.

   The three-level laser has a number of advantages, one of which is that a range of exciting frequencies can be used to stimulate the electrons initially. This makes the pumping process more efficient, leading to a higher population inversion. In this case the high-intensity light source has a higher frequency than that which the laser is to emit. Electrons are stimulated into the higher energy, short-lifetime states of the 'pump band' and then de-excite to the narrow intermediate level. The lifetime of the electrons in this narrow, intermediate energy level is much longer, being typically milliseconds, and this allows the population inversion to be maintained. The electrons are then stimulated to de-exite to the lowest level with the emission of a photon.

12.3.6   Four-level laser

Are there further refinements of the electronic structure which can lead to improved performance of the laser?

A further refinement is the four-level laser (Fig. 12.8) which has two lower energy levels. This results in emptying of the energy level E_2 into energy level E_1, and hence an even larger population inversion between levels E_2 and E_3 between which the stimulated emission of light occurs.

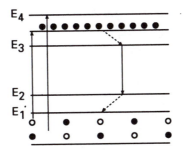

Fig. 12.8 Electron energy diagram for a four-level laser.

12.3.7   Laser materials

What materials are used in solid-state lasers?

A wide range of materials are currently used in the construction of solid-state devices for the generation of laser light. These include:

- ruby (Al_2O_3, Cr^{3+} doped)
- glass (Nd^{3+} glass or YAG)
- gases (helium, neon)
- vapours (mercury, cadmium)
- CO_2
- liquids (dye lasers)
- semiconductors (gallium arsenide and related III–V semiconductors).

12.3.8   Semiconductor lasers (laser diodes)

How does the semiconductor laser work?

The semiconductor laser can be described fairly easily in terms of the electronic properties of materials discussed earlier. The cavity (Fig. 12.9) consists of pieces of doped n- and p-type semiconductors in the form of a single p-n junction which we have described above. In such a junction, the electron band structure is distorted under forward biasing, as shown in Fig. 12.4. This gives a light-emitting diode, but by arranging that the light emission only occurs when stimulated by photons of the same energy, laser action can be set up.

The stimulated emission only occurs where there is a high population of excited electrons created by 'pumping'. Pumping occurs by direct injection of electrons into the depletion layer where the electrons remain in a metastable state until they are stimulated to return to the valence band by the passage of a photon of the correct energy. The transition only occurs in the depletion layer.

12.3.9   Types of semiconductor-junction lasers

What different types of semiconductor laser are there?

When the two parts of the junction are made from the same material the device is known as a homojunction laser. When the p and n parts of the junction are made from different materials the device is known as a heterojunction laser.

The heterojunction lasers have some advantages if constructed in such a way that the refractive index of the active region is larger than the surrounding regions. This produces an optical waveguide which means that the laser action is confined to the junction layer and no heat is absorbed by the remaining material. The heterojunction laser has higher efficiency because the laser light confined to the junction layer also helps to stimulate more electrons to de-excite.

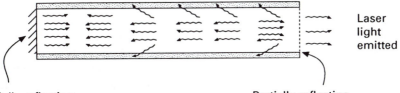

Laser
light
emitted

Fully reflecting   Partially reflecting

Fig. 12.9 Stimulated emission of radiation in the semiconducting material of a laser. The two ends contain parallel mirrored surfaces, and these, combined with the elongated junction region ensure low beam divergence.

12.3.10 Applications of semiconductor lasers

Where do semiconductor lasers find specialized applications that differ from the larger high-power lasers?

Helium–neon lasers which have ultra-low beam divergence are used for alignment purposes. Highly focused pulsed laser beams can be used for drilling narrow holes. Optical disks for audio and digital recording applications need to be encoded and read using a laser beam. The laser diode is also used in digital audio and video disk players to read the information encoded on tracks moulded onto the disks. Lasers are also used as a cutting tool in laser surgery.

In telecommunications the very high frequency of the laser light (10^{14} Hz) enables intensity to be rapidly altered to encode complex signals. The emitted light signal can be modulated at high frequencies by the voltage applied to the semiconductor [7, 8]. This allows the possibility of optical communications at high frequencies and over long distances.

12.4 FIBRE OPTICS AND TELECOMMUNICATIONS

What are the advantages of optical communications over conventional electrical communications?

In many cases light pipes are needed as waveguides to ensure that emission of light from a source reaches the appropriate detector. The preferred method is the use of optical fibres. These are transparent materials fabricated with a high length-to-diameter ratio to ensure that light travelling down them is totally internally reflected from the side walls and therefore that very little energy is lost.

Since the emission of light as a result of laser action is dependent on the applied electric field, or on the injected current, it becomes possible with a semiconductor laser to modulate the optical signal through control of the electric field or the injected current. This means that optical communication can be achieved in this way [9], and such communication has distinct advantages over conventional electrical communication in terms of speed and distance over which communication can be maintained.

Optical fibres are made from glass or plastic ('plexiglass' or 'perspex'). These are solid tubes which act as waveguides for light. High-quality optical fibres are made from glass as the attenuation is lower than for plastic fibres. The high attenuation of plastic fibres means that they are only used in a limited number of applications usually for shorter distance communications. The glass optical fibres are thinner than a human hair and consist of a core region with larger refractive index and an outer cladding region with lower refractive index. The refractive index of the material is changed using dopants such as germanium dioxide, phosphoric oxide or boric oxide. This arrangement confines the light to the core region. Core diameters are typically 1–100 µm in diameter, while the cladding is typically 100–300 µm in diameter.

Optical fibres can be made in single- and multi-mode forms. The single-mode fibre is thinner in diameter with a 1–10 µm core. The refractive index of the outer cladding layer is 0.1–0.3% lower than the core region and there is usually a discontinuous change in refractive index from the core to the cladding. The multi-mode fibres have a core diameter of typically 40–100 µm and a refractive index change of 0.8–3.0% from the core to the cladding. These multi-mode fibres can have either discontinuous or graded refractive index change from core to cladding. In these multi-mode fibres multiple electromagnetic field configurations can propagate down the fibre simultaneously [10].

12.5 LIQUID-CRYSTAL DISPLAYS

How do liquid-crystal displays work?

Liquid crystals combine the fluidity of liquids with the orientational (anisotropic) properties of solid crystals. The viscosity of the liquid is similar to machine oil. Nematic liquid crystals, which are often used for displays, are composed of elongated organic molecules which align in a preferred direction [11]. In displays, the molecules of the liquid must be alignable in some way, and this is usually achieved by an electric field. Polarized light is employed through the use of polarized filters in the liquid crystal display (LCD) screen. The back surface of the LCD display screen has a mirror to reflect incident light. The orientation of the molecules in the liquid crystals changes the direction of polarization of light (Fig. 12.10). This can then be used to produce bright or dark regions on the surface of the display screen.

Liquid-crystal displays are now widely used and are replacing light-emitting diode (LED) displays in many but not all applications. They are increasingly popular as flat-panel displays for computer screens where they are rapidly replacing the rather bulky cathode-ray tubes (CRTs). In addition to their small size one of the other advantages of liquid-crystal displays is their low power consumption. The LCD modifies the ambient light instead of generating its

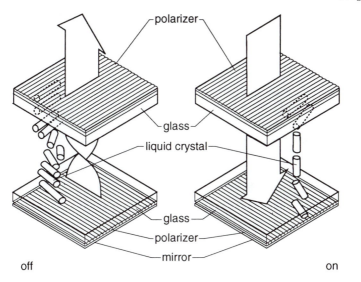

off   on

Fig. 12.10 Arrangement of a liquid-crystal display showing the liquid-crystal elongated organic molecules, polarizing filters, mirror and glass substrates.

own light as the LED does. Some LCD displays can operate for more than a year on small batteries and so these types of display are favoured for pocket calculators.

Another type of display involving the use of ferroelectric materials instead of liquid crystals is also currently under investigation.

REFERENCES

1. Forrest, S. R. (1986) *IEEE Spectrum*, **23** (5), 76.
2. Blakemore, T. S. (1992) *Gallium arsenide*, Key papers in Physics series, American Physical Society, New York.
3. Stillman, G. E., Robbins, V. M. and Tabatobaie, N. (1984) *IEEE Trans. Electron. Devices*, **31** (11), 1643.
4. Wu, M. C. and Chen, C. W. (1992) *J. Electr. Mater.*, **21**, 977.
5. Li, J. W., Su, Y. K. and Yokoyama, M. (1992) *J. Electr. Mater.*, **21**, 659.
6. Goodhue, W. D. *et al.* (1990) *J. Electr. Mater.*, **19**, 463.
7. Bell, T. E. (1986) *IEEE Spectrum*, **23** (8), 38.
8. Kobayashi, S. (1984) *IEEE Spectrum*, **21** (5), 26.
9. Guterl F. and Zorpette, G. (1985) *IEEE Spectrum*, **22** (8), 30.
10. Kimura, T. (1988) *Lightwave Technol.*, **6**, 611.
11. Kahn, F. J. (1982) *Physics Today*, **35**, 66.

FURTHER READING

Gunshor, R., Nurmikko, A. and Kobayashi, M. (1992) *Physics World*, **5** (3), 46.
Inagaki, T. (1989) Hologram lenses lead to compact scanners, *IEEE Spectrum*, **26** (3), 39.

Skromme, B. J., Bose, S. S. and Stillman, G. E. (1986) New shallow acceptor levels in GaAs, *J. Electr. Mater.*, **15**, 345.

Journals

Journal of Electronic Materials (published jointly by the IEEE and TMS)
IEEE Transactions on Electron Devices

13

Quantum Electronics – Superconducting Materials

The main objective of this chapter is to give an overview of superconductivity which includes a description of the basic observations of the phenomenon and an indication of the principal applications. We discuss the emergence of super-conductivity in certain materials and how these materials are used in three main groups of applications: superconducting solenoids, superconducting magnetometers (SQUIDs) and superconducting logic devices. Both flux pinning by a superconductor and the Meissner effect are explained, together with the differences between Type I and Type II superconductors. The onset of superconductivity is discussed as a discontinuous reduction in conductivity to a state with zero DC resistance. It is shown that the resistanceless state is insufficient to explain the Meissner effect in which magnetic flux is completely excluded from the bulk of a superconducting material. Conditions for establishing the presence of superconductivity are given.

13.1 QUANTUM EFFECTS IN ELECTRICAL CONDUCTIVITY

Which are the areas in which quantum effects radically alter the electronic properties of materials on the bulk scale?

By quantum electronics we mean any electronic behaviour where quantum effects make their presence felt. Of course we can argue that all electronics is quantum electronics. So here we are interested only in those effects where the quantum description is very different from the classical description. Super-conductivity, the apparently complete loss of resistivity in some materials at low temperatures, is a quantum phenomenon with no adequate explanation on the classical scale. It results in the passage of an electric current in a material without a potential difference to drive it. We shall consider how this can be used for particular functions. In fact there are three main areas of use: super-conducting magnets, superconducting detectors (SQUIDs), and superconducting devices for electronics. Other applications include superconducting light detectors.

13.1.1 Reduction in resistance on cooling

What happens to the resistance of a material when it goes superconducting?

Superconductivity was first discovered by Kamerlingh Onnes in mercury in the year 1911 [1]. The metal exhibited a very rapid reduction in resistivity at 4.2 K, as shown in Fig. 13.1. The fact that the resistivity is very low (typically $< 10^{-25}\,\Omega\,\mathrm{m}$) is a necessary but insufficient condition for the existence of superconductivity. Naturally, we expect a reduction of resistivity as a result of the reduction of lattice vibrations which in turn reduces the amount of phonon scattering of electrons. However, impurity scattering should continue even at the lowest temperatures and so a residual resistance to the motion of electrons is expected on the basis of classical theory. The reduction of resistivity observed in normal conductors is a continuous process, whereas in superconducting materials there is a sharp phase transition between the normal and superconducting states. Superconductivity therefore involves something else.

13.1.2 Superconductivity

How do we recognize the existence of superconductivity?

The two essential criteria for determining the existence of superconductivity are:

1. complete disappearance of resistivity below a critical temperature T_c;
2. exclusion of magnetic flux from the superconductor – the Meissner effect.

   The first condition means that an electric current continues to flow even in

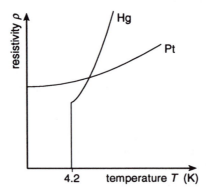

Fig. 13.1 Variation of resistance with temperature in mercury and platinum. The superconducting transition in mercury begins at about 4.26 K as the material is cooled. In platinum, which exhibits conventional behaviour, the contributions to resistivity include impurity or defect scattering, which is represented by the resistivity $\rho(0)$ at $T = 0$, and phonon scattering, which is temperature dependent and is represented by $\rho(T) - \rho(0)$.

the absence of an applied voltage. A further observation is that there exists a critical applied field H_c above which the material is driven 'normal'. This critical field is temperature dependent and becomes zero at the critical temperature T_c.

13.1.3 Nature of the superconducting transition

What are the principal changes in material properties that are observed as a result of superconductivity?

The onset of superconductivity leads to some quite dramatic changes in properties. The following list indicates some of the major observations.

1. There is a discontinuous reduction in resistivity.
2. In the superconducting phase there is a different conduction mechanism.
3. AC (eddy current) losses remain, even though there are no DC losses.
4. Resistivity is more than 13 orders of magnitude lower than in the best high-purity annealed copper ($10^{-25}\,\Omega\,m$ in a superconductor compared with $10^{-12}\,\Omega\,m$ in high-purity copper and $10^{-8}\,\Omega\,m$ in normal copper used for conducting wires).
5. There is a critical temperature T_c above which the material reverts to normal conducting behaviour.
6. There is a critical field H_c above which the material reverts to normal conducting behaviour.
7. There is a critical current density above which the material reverts to normal conducting behaviour.

The critical field strength and the critical temperature are related by an

Table 13.1 Critical temperatures and critical fields for superconducting elements

Material	Critical temperature T_c (K)	Critical field H_0 ($kA\,m^{-1}$)
Aluminium	1.2	7.9
Cadmium	0.5	2.4
Indium	3.4	22
Lead	7.2	64
Mercury	4.2	33
Niobium	9.3	Type II
Tantalum	4.5	66
Tin	3.7	24
Zinc	0.9	4.2
Zirconium	0.8	3.7

equation of the form,

$$H_c = H_0(1 - T^2/T_c^2)$$

where H_0 is the critical field strength at 0 K.

Below the transition temperature the material has zero resistance unless the current passing through it becomes too large (i.e. reaches the critical current density J_c) in which case the material returns to its normal resistive state.

13.2   THEORIES OF SUPERCONDUCTIVITY

How can superconductivity be explained?

Over the years there have been three main theories of superconductivity, the phenomenological two-fluid model of Gorter and Casimir [2], the mathematical electrodynamic theory of London and London [3] and the paired-electron theory of Bardeen, Cooper and Schrieffer [4], which is a 'first principles' microscopic theory of superconductivity. Here we will discuss only explanations arising from the last of these.

In normal metals, individual electrons are scattered by impurities and by phonons. This causes resistance in the material to the passage of electrons which results in macroscopic resistivity, as described in Chapter 2. Under certain conditions however, the electrons at energies near the Fermi level, which are the electrons which contribute to the conductivity, can couple together. This changes their energies and leads to a superconducting energy gap. These electrons then move throughout the solid without scattering as a coherent group. Under these conditions the phonons and impurities are too weak to scatter the electrons and so the electric current once started moves through the material without experiencing resistance. In the Bardeen–Cooper–Schrieffer (BCS) theory the electrons are paired with opposite spins and opposite wave vectors which results in no net spin and no net momentum. These 'Cooper pairs' are not scattered by the normal mechanisms because their combined wave vector k is zero, corresponding to an infinite wavelength.

13.2.1   Conditions for superconductivity to occur

Why does superconductivity occur only at low temperaures and only in certain materials?

For superconductivity to occur in a material there must be strong phonon interactions between the conduction electrons and the lattice. The conduction electrons can then be coupled together indirectly, via this phonon interaction, into superconducting 'Cooper pairs'. The energy change associated with this coupling is typically in the range 10^{-3} eV, and this leads to a superconducting energy gap of this size at the Fermi level.

If the temperature is raised, lattice vibrations begin to disrupt the interaction

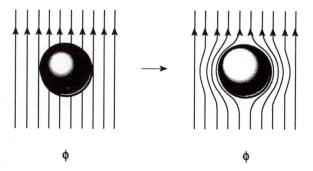

Fig. 13.2 Diagram showing flux exclusion from the superconducting material once it has undergone a superconducting transition.

between the coupled electrons until at a critical temperature T_c the lattice vibrations completely destroy the coupling. The superconducting electrons then revert to their normal state. The thermal energy needed to decouple the electrons is of the same order of magnitude as the superconducting energy gap. For an energy gap of 10^{-3} eV this gives a critical temperature of a few degrees Kelvin, which is typical of many superconducting materials as shown in Table 13.1.

Therefore it can be seen that the superconducting state is eliminated at higher temperatures. Furthermore, since strong interaction between the conduction electrons and the lattice is a prerequisite for superconductivity, it follows that metals such as copper, silver and gold, which are good electrical conductors and therefore have weak interactions between the conduction electrons and the lattice, do not exhibit superconductivity.

13.2.2   The Meissner effect

How does a superconductor respond to an external magnetic field?

When a magnetic field is applied to a superconductor it has the effect of destroying the superconductivity above the critical field H_c. This occurs once the field strength is large enough to elevate the energy levels associated with the superconducting state above those of the normal state. In this case the material will again revert to its normal resistive state. At lower field strengths the superconducting material may completely exclude the magnetic field, a phenomenon known as the Meissner effect [5]. This is shown schematically in Fig. 13.2.

In this case the superconducting electrons set up circulating currents at the surface of the material which counteract the applied field and so cancel the field exactly to zero within the material. In some superconductors, known as Type II superconductors, there also exists a state at intermediate field strengths in which the field penetrates local regions of the superconductor. This field penetration is through flux tubes of normal material embedded in a matrix of superconducting material. This state is known as the 'mixed' or 'vortex' state.

13.2.3   Type I and Type II superconductors

Is the transition to the superconducting state continuous or discontinuous when a magnetic field is applied?

Superconductors fall into two categories based on the mechanism of the transition from superconducting to normal state in the presence of a magnetic field. These are known as Type I and Type II superconductors.

Type I superconductors exhibit only two phases: normal and superconducting. The transition between these phases is very sharp and occurs at a particular critical field strength H_c. Most pure metals are Type I superconductors.

Type II superconductors exhibit an intermediate phase known as the 'mixed' or 'vortex' state. This state exists at field strengths between H_{c1}, the critical field for transition from the flux free (Meissner) state to the mixed state, and H_{c2}, the critical field for transition from the mixed state to the normal state. In this intermediate state the material consists of a honeycomb structure of normal material within a superconducting matrix. Lines of magnetic flux which cannot penetrate the superconducting material because of the Meissner effect, can enter through the normal material. Each 'vortex' of normal material carries one flux quantum of 2.07×10^{-15} Wb.

Most alloys are Type II superconductors. The matrix of flux vortices in the material of a Type II superconductor in its intermediate mixed state is shown in Fig. 13.3. The tubular regions are where the magnetic flux quanta emerge perpendicular to the surface through the normal material. The remaining regions are the superconducting material.

The usefulness of Type II superconductors arises from the extension of the range of magnetic field strength over which superconductivity can occur.

13.2.4   Flux pinning and flux exclusion

Is a superconductor merely a perfect conductor or is there a greater significance to the Meissner effect?

For some years after the discovery of superconductors, it was assumed that the behaviour of such a material in the presence of a magnetic field would be as shown in Fig. 13.4. That is to say, the supercurrents would prevent any change in the flux passing through the material because by the Faraday–Lenz law of electromagnetic induction they would set up an induced current. In a perfect conductor this induced current should produce an opposing flux change which exactly counteracts the flux change producing it.

In fact the situation is quite different as discovered by Meissner and Ochsenfeld [5]. Inside a superconductor in its superconducting state we find that the magnetic flux Φ is zero, except for a thin boundary layer at the surface (Fig. 13.5). (We must note immediately that this condition is not exactly fulfilled in the mixed state because of the presence of normal material). The Meissner effect does

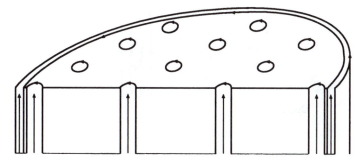

Fig. 13.3 Flux lines in a superconductor in its mixed state. The tubular regions are normal material with magnetic flux penetration within a matrix of superconducting material.

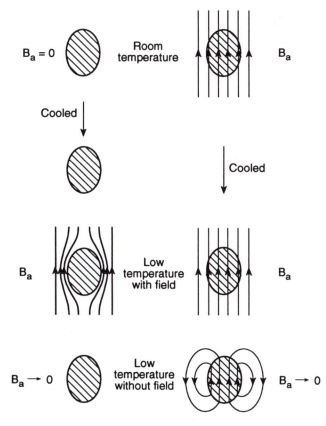

Fig. 13.4 Diagram showing the expected behaviour of a 'perfect conductor' in the presence of a field.

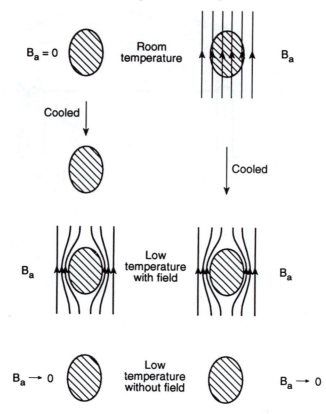

Fig. 13.5 Equivalent behaviour of a superconductor (compare with Fig. 13.4), which exhibits the Meissner effect or flux exclusion.

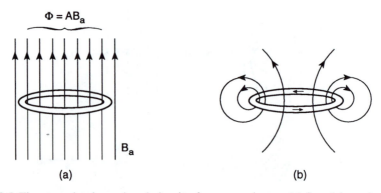

Fig. 13.6 Flux trapping by a closed circuit of superconductor: (a) flux driven through a ring of normal material by an applied field; (b) flux trapped by the ring when the ring goes superconducting and the field is subsequently removed.

Table 13.2   Penetration depths in superconductors

Material	Penetration depth λ $(10^{-8}\,m)$
Aluminium	1.6
Cadmium	11
Lead	3.7
Niobium	3.9
Tin	3.4

however demonstrate that the superconducting state is something more than just a state with perfect conductivity, since the exclusion of flux is an additional property that a merely resistanceless material would not possess.

13.2.5   Surface currents and the Meissner effect

How can we explain the Meissner effect?

When a superconductor is cooled in a magnetic field, persistent currents arise on the surface of the material at the critical temperature and these circulate so as to exactly cancel the flux density inside. The surface supercurrents are determined only by the strength of the external prevailing magnetic field.

The emergence of the surface currents when a material is cooled through its superconducting transition lies beyond the concept of 'perfect' conductivity. In order not to get an infinite current density, these surface currents must exist over a finite depth. In fact the surface currents decay exponentially with depth, and this means that the magnetic field does penetrate at the surface of the super-conductor to some extent. This is expressed by the penetration depth λ. Typical penetration depths in superconductors are 10^{-8} m. Some values are shown in Table 13.2

13.2.6   Flux trapping in a superconducting circuit

What happens when a magnetic flux passes through the middle of a circuit of superconducting material?

We know that the Meissner effect leads to the exclusion of magnetic flux passing through the body of a superconducting material. A related phenomenon is flux trapping through a superconducting circuit, as shown in Fig. 13.6.

If we have a toroid of superconductor with a flux Φ passing through the circuit as shown, then the flux passing through the circuit cannot change. The reason lies again in the Faraday–Lenz law of electromagnetic induction which states that any change in flux linking a circuit $d\Phi/dt$ sets up a counteracting current in the circuit. In the case of a superconductor, this current produces a flux which exactly counteracts the flux change producing it. This is the principle behind the SQUID (superconducting quantum interference device), which is used for measuring changes in magnetic flux to extremely high resolution.

13.3   RECENT DEVELOPMENTS IN HIGH-TEMPERATURE SUPERCONDUCTORS

How have the ceramic superconductors altered the perspective for applications of superconducting materials?

General awareness of superconductors was raised by the discovery of materials which are superconducting at temperatures above 77 K. Until early 1986, the highest known critical temperature for superconductors was 23.2 K in Nb_3Ge, and in fact over the previous 75 years the critical temperatures of superconductors had been raised only very gradually.

Rapid developments in superconductivity began with Bednorz and Muller [6] who discovered that La-Ba-Cu-O is superconducting at 30 K. Within a year, an alloy of Y-Ba-Cu-O had been found with a T_c of 95 K, and within two years an alloy of Tl-Ba-Ca-Cu-O with T_c of 125 K had been discovered. These discoveries were of great scientific interest because they have raised the possibility of finding a room-temperature superconductor. Also there has been renewed interest in research in two areas, (i) development of new superconductors and (ii) theoretical explanation of the conduction mechanism in these superconductors.

There were also some drawbacks to these materials however, because they are extremely brittle after their final heat treatment and they unfortunately have quite low critical current density before going 'normal'. There will certainly be a niche for these materials, but it seems likely that, in the near future at least, they will be used principally in low power or small electronic device applications. For example thin films of high-temperature superconductors are already commercially available for SQUIDS and microwave devices [7] both of which are low-power applications.

Perhaps the renewed interest in the field of superconductivity will lead to the discovery of more practical high-temperature superconductors, and if so this will be a significant technological advance. There are now collaborative research and development efforts between major US corporations, universities and National Laboratories in order to explore possible applications of these materials in superconducting electronics.

13.3.1   Critical current density and critical field strength

Are critical current density and critical field strength related?

In the new ceramic superconductors the low critical current densities are still problematic. In bulk specimens of La-Ba-Cu-O for example, the value is $1.5\,kA\,cm^{-2}$, whereas in a material such as niobium-tin it is typically $10\,MA\,cm^{-2}$.

However in these ceramic superconductors the critical field strength H_{c2} is extremely high and has been estimated at $H_c = 3 \times 10^6$ Oe (0.24×10^9 A m^{-1}) which is too high to measure under present capabilities. An interesting conclusion from this result is that there does not appear to be a connection between the critical fields and the critical current densities in these materials.

13.4 APPLICATIONS OF SUPERCONDUCTORS

What are the major technological applications of these materials?

The major applications of superconductors can be categorized into three main areas:

1. generation of high magnetic fields using superconducting solenoids, for example in magnetic resonance imaging (MRI);
2. high-resolution detection of magnetic flux using superconducting quantum interference device magnetometers (SQUIDs);
3. small, low-power electronic devices based mostly on the Josephson effect.

The discovery of ceramic superconductors with critical temperatures above the boiling point of liquid nitrogen has affected each of the three main areas of applications. A review of applications of superconductors can be found in the text by Ruggiero and Rudman [8] which consists of a collection of review articles each written by a leading expert on a particular device application. Although the level of the treatment is rather too advanced and detailed for the needs of the present book, the work does provide an excellent source of reference for recent developments in superconducting devices.

On a more introductory level, the basic science and engineering of superconductivity has been dealt with by Orlando and Delin [9], in which superconducting junctions and devices are discussed. Another useful guide to engineering applications of superconductivity is the work by Doss [10], which although ultimately directed towards ceramic superconductors, nevertheless also contains a detailed general discussion of applications of superconductivity.

13.4.1 Superconducting solenoids and magnets

How are superconducting materials used to produce high magnetic fields?

One of the principal applications of superconductors is in the generation of high-intensity, high-stability magnetic fields for both scientific investigations and medical applications such as magnetic resonance imaging (MRI). In this case superconducting wires need to be fabricated to form the coils of the solenoid. These wires usually consist of niobium-titanium or less often niobium-tin in a matrix of copper which is extruded into a wire about 0.5 mm in diameter. These multifilament wires are shown in cross-section in Fig. 13.7.

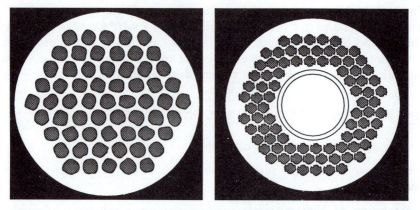

Fig. 13.7 Cross-section of a superconducting wire; the dark regions are the superconducting material, and the light regions are a matrix of normal conducting material such as copper. Two configurations are popular; in the left-hand diagram superconducting filaments are distributed throughout the matrix, in the right-hand one a core of normal material is surrounded by superconducting filaments.

In magnetic resonance imaging for medical diagnostics the magnetic moment of the nucleus can be detected through nuclear magnetic resonance. The resonant frequency of a particular nucleus is dependent on its mass, nuclear magnetic moment and the field strength it is subjected to. The resonance can be caused by using an rf coil with an adjustable excitation frequency and detected by using a pick-up coil.

A field gradient is normally used to determine the spatial locations of the nuclei. The measured resonance frequency indicates the field strength which a particular nucleus experiences. From a three-dimensional map of the field strength over a given volume, the location of the nucleus within that volume can be found. In this application it is essential to have a strong magnetic field which is stable and precisely controllable over the working volume. A high field strength is advantageous because it gives a higher resonant frequency and therefore a stronger signal to noise ratio. In this type of instrument the magnet is therefore a critical component.

Superconducting magnet systems which can generate magnetic flux densities of up to $B = 15$ tesla ($H = 12 \times 10^6 \, A \, m^{-1}$, or 150 kOe in free space) have been available for many years. The wires which are used to make the coils of the superconducting solenoid can carry much higher current densities than conventional conductors. For example in niobium-tin the critical current density, which provides an upper limit to the current density the material can sustain before making a transition to the normal or resistive state, is typically $10 \, MA \, cm^{-2}$ ($10^{11} \, A \, m^{-2}$).

13.4.2 Superconducting magnetometers

How can superconducting flux pinning be used to develop high sensitivity magnetometers?

In a superconductor–insulator–normal metal (SIN) junction, quantum mechanical tunnelling of the electrons across the energy barrier presented by the insulator can occur, provided the insulating layer is sufficiently thin. This can be compared with tunnelling of electrons into a finite potential barrier as discussed in section 4.3.2. Single-electron tunnelling was first demonstrated by Giaever [11]. Later Josephson [12] predicted that in a superconductor–insulator–superconductor (SIS) junction, quantum mechanical tunnelling of Cooper pairs could occur with a higher probability than had been generally recognized before. This effect was demonstrated experimentally by Anderson and Rowell [13].

When two such SIS or Josephson junctions are connected in parallel there is a quantum interference between the electron wave functions. This leads to an oscillatory variation in the voltage across the device with magnetic flux linking the circuit. This can be used as a high-resolution flux counting device known as a superconducting quantum interference device (SQUID). SQUID magnetometers can be used to measure changes in flux down to a micro flux quantum of 10^{-21} Wb. The flux quantum is 2.07×10^{-15} Wb. They are now used in medical diagnostics, mineral surveying, submarine detection, motion detection, materials evaluation and scientific measurements [14].

If we consider first a superconducting circuit enclosing a flux Φ, as shown in Fig. 13.8, then there will be an outer supercurrent I_0 which exactly compen-

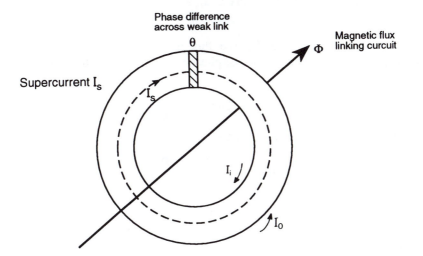

Fig. 13.8 Arrangement of a circuit of a superconductor with a weak link to form a Josephson junction.

sates the external field in order to prevent flux passing through the super-conducting material. This outer current does not depend on the field exposure history, it depends only on the external field strength, that is it depends entirely on the prevailing conditions.

In order to maintain the flux through the hole in the centre of the toroid there must be an inner current I_i which depends on the field exposure history of the toroid. This current is determined by the amount of flux that was passing through the circuit at the time the material became superconducting.

The relationship between the flux density threading the superconducting circuit and the flux density of the applied field is

$$B = B_a + \frac{LI_s}{A}$$

where B_a is the flux density of the applied field, I_s is the total superconducting current, L is the inductance of the ring, and A is its cross-sectional area. The supercurrent I_s is related to the critical current I_c which is determined by the properties of the weak link,

$$I_s = I_c \sin \theta$$

where θ is the phase difference of the electron pair wave function across the weak link. Therefore, we can write the relationship between the flux densities as

$$B = B_a + \frac{L}{A} I_c \sin \theta.$$

If we have a completely superconducting ring, then the flux through the ring must be an integral number of flux quanta

$$\Phi = N\Phi_0$$

where N is an integer and Φ_0 is a flux quantum. With the weak link or Josephson junction in the circuit, the phase angle θ depends on the flux Φ in the following way,

$$\theta = 2\pi N - 2\pi \frac{\Phi}{\Phi_0}$$

and since N is an integer we must have,

$$\sin \theta = \sin 2\pi \left(N - \frac{\Phi}{\Phi_0} \right)$$

$$= -\sin \left(2\pi \frac{\Phi}{\Phi_0} \right)$$

so that

$$B = B_a - \frac{L}{A} I_c \sin \left(\frac{2\pi\Phi}{\Phi_0} \right).$$

This means that each time Φ equals an integral multiple of 2.07×10^{-15} Wb, the flux density in the ring is equal to the flux density of the applied field. However at intermediate values of flux a supercurrent flows in the superconductor. This supercurrent is determined by the flux entering the ring and it can be measured. If a loop of wire is wound on to the superconducting ring, then the voltage induced in the loop is a periodic function of the flux linking the circuit. This can therefore be used to count the changes in flux quanta.

13.4.3 Principles of operation of a SQUID

How does a SQUID count the flux changes?

If we now consider the situation depicted in Fig. 13.9, a current I flows through the two paths of the interferometer device. We assume that the device is symmetric so that $I/2$ flows through each arm of the device. Assuming $I < 2I_c$ then we will have a phase angle across each weak link as described above. Now suppose a flux Φ is introduced into the loop. This will cause a superconducting current I_s, which will add to the existing current in one arm but subtract from

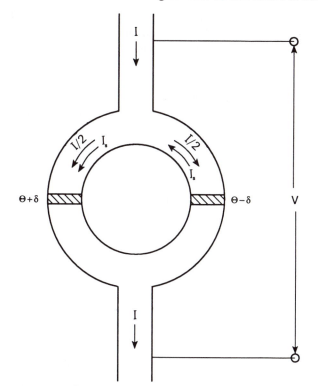

Fig. 13.9 Electrical circuit showing connection of two Josephson junctions to form a SQUID.

the current in the other arm. The phase angles across the two weak links will be $\theta + \delta$ and $\theta - \delta$ so that

$$\tfrac{1}{2}I + I_s = I_c \sin(\theta + \delta)$$
$$\tfrac{1}{2}I - I_s = I_c \sin(\theta - \delta).$$

Summing these currents gives,

$$I = I_c\{\sin(\theta + \delta) + \sin(\theta - \delta)\}$$
$$= 2I_c \sin\theta \cos\delta$$

and the total change of phase across the two weak links is 2δ which must equal $2\pi N - 2\pi\Phi/\Phi_0$, so that,

$$\delta = \pi N - \frac{\pi\Phi}{\Phi_0}$$

where N is an integer as before. Therefore

$$I = 2I_c \cos(\pi N) \cos\left(\frac{\Phi\pi}{\Phi_0}\right) \sin\theta.$$

Therefore as the flux is increased the current varies periodically with flux, and hence so does the voltage V across the device, as shown in Fig. 13.10. The period is one flux quantum Φ_0, so the device can be used to count flux quanta, and in fact its resolution can be as low as 10^{-6} of a flux quantum (depending on the signal to noise ratio). Note however that in practice the current never changes direction, as the above equation seems to imply. This is because the electron pairs adjust their phase θ to ensure that the current I is always flowing in the same direction.

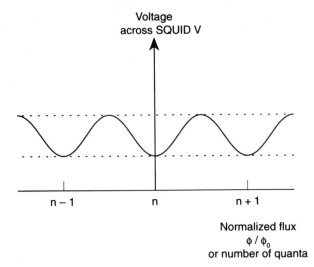

Fig. 13.10 Variation of the voltage across a SQUID with the number of flux quanta in the external field attempting to pass through the ring.

Note that the device measures changes in magnetic field rather than the absolute value of magnetic field. The device therefore can be used to count the number of flux quanta which enter the ring.

A new type of SQUID magnetometer without Josephson junctions was proposed by Fink *et al.* [15]. In this device the supercurrents in the two parallel branches of the device result in a critical current which varies in an oscillatory way with the magnetic flux linking the circuit. Recent work by Moschalkov *et al.* [16] has shown that this quantum interference does occur in a superconducting loop of length no greater than the Cooper pair coherence length, which is typically a few micrometres and is much smaller than the length of loop used in a conventional SQUID.

This leads to cyclic variation in the voltage across the loop as a function of the magnetic flux linking the loop. The mechanism of the interference in these mesoscopic loops is quite distinct from the effect observed in Josephson junction devices. However there are some similarities in performance. Regions of low density of Cooper pairs, which are analogous to the weak links in the Josephson junction devices, occur periodically in these mesoscopic loops at half-integer flux quanta as a result of the effects of shielding and transport currents in the loop. The phase coherence of electrons in the loop ensures that when the 'pseudo-weak link' appears in one branch of the loop it must simultaneously occur in the other branch.

There is a long-term prospect for the development of electronic circuit elements based on the critical current oscillations in these mesoscopic loops. However there are significant practical problems to be overcome. Specifically, in order to observe the effect at present, temperatures below 1 K are required with voltages of less than 0.1 mV. The response of the device is also very sensitive to changes in sample configuration. For example the interference pattern can be changed even by moving a single impurity in the device. This means that considerable development work is required to ascertain whether such devices can be made practicable in future.

13.4.4 Superconducting electronic devices

Can superconducting switching effects be used to construct logic devices for computers?

We consider now 'electronic' or low-current applications as distinct from 'electrical' or high-current applications of superconductors. In these applications the electronic characteristics of the superconducting junctions perform traditional electronic functions, although in many cases with improved performance over conventional (i.e. non-superconducting) materials. For example superconducting devices are often faster, or can operate at higher frequencies, or have lower losses than conventional materials.

Along with the change in resistivity of superconductors other equally dramatic changes can occur in the electronic properties, including a change in the density of states which arise because of the superconducting energy gap. Single-electron tunnelling [11] occurs when electrons tunnel through a thin insulating layer between a superconducting material and a normal material known as an SIN junction. A related phenomenon is the Josephson effect [12] in which a coupled pair of electrons (a Cooper pair) passes from one superconducting region to another through a thin insulating layer known as an SIS junction. This effect can be used to construct devices which can change from one electrical state to another at a very fast rate.

This offers the opportunity for constructing logic devices and small computer circuits which operate very rapidly, with switching speeds that can be up to 1000 times faster than conventional silicon devices. Efforts to construct a computer on this basis have encountered practical difficulties and have also suffered from the competition from the rapidly developing III–V semiconductors which can also be used for high-speed devices.

The main electronic device and circuit applications of superconductors are broadly in the following areas:

- radio frequency and microwave devices (such as millimeter wave detectors and mixers, filters, resonators and phase shifters)
- high-speed digital logic devices and circuits
- low-noise, high-frequency analogue devices (for example components for high-speed oscilloscopes)
- thin-film devices
- hybrid superconductor–semiconductor devices and circuits
- optical detectors.

These various applications of superconductors have been discussed in detail by Van Duzer and Turner [17], and more recently by Ruggiero and Rudman [8] and by Doss [10].

Ralston, Kastner, Gallagher and Batlogg [18] have given a review of some of the latest developments in low-current device applications of superconductors. In work performed under an industry–university consortium for superconducting electronics, four main areas of interest have been explored:

- materials and processing
- superconducting junctions
- networks
- circuits.

In materials and processing it was necessary to develop large-area substrates in order to prepare the thin films. This meant that both the lattice parameters and the thermal expansion coefficient of the substrate should match those of the thin film. The main candidate materials that have been investigated are, neodymium gallate ($NdGaO_3$) and lanthanum aluminate ($LaAlO_3$), magnesium oxide (MgO) and cubic zinc oxide (ZnO_2). The deposition of the thin films of superconductors onto the substrates can proceed by one of several methods including sputtering, co-evaporation in vacuum, laser ablation, molecular-beam epitaxy (MBE), and organometallic deposition.

Materials which are receiving much attention for thin film devices are the yttrium barium copper oxide material ($YBa_2Cu_3O_x$ or 'YBCO'), thallium barium calcium copper oxide (TlBaCaCuO or 'TBCCO') and bismuth strontium calcium copper oxide (BiSrCaCuO or 'BSCCO'). Another material which is being investigated is barium potassium bismuthate ([BaK]BiO_3) which has a very long coherence length, that is the range over which electron states are correlated. In addition its electronic properties are isotropic. These characteristics make the material a good candidate for devices based on tunnelling effects and junctions, despite its comparatively low critical temperature of 30 K.

In developing junction devices using other ceramic superconductors, there are some serious challenges to be overcome which arise because of the anisotropy and short coherence length. However a prototype junction device using thin films of ceramic superconductor has been developed and this is shown in Fig. 13.11. This is a Josephson junction device made from YBCO thin films which are separated by a layer of praesodymium barium copper oxide $PrBa_2Cu_3O_x$. The films were grown with the copper oxide planes parallel to the substrate surface, giving the highest critical current density in this direction. The step structure of the films is arranged to ensure that the weak link is also in the direction of maximum coherence length. This geometry therefore optimizes the performance of the device.

In networks, the superconductors can be used to interconnect between semiconductors. The superconducting thin films have low resistivity at microwave frequencies and so low-loss compact microwave filters can be made from them. For example at 77 K and 4 GHz a YBCO filter has losses which amount to only 25% of an equivalent filter fabricated from silver. The high-performance superconducting microwave filters can have Q values of better than 1000, compared with Q values of 250 in filters fabricated from normal metals operating under similar conditions.

ultra-low-noise SQUIDs can be made using a three-layer process employing niobium–aluminum oxide–niobium films. These are produced by sputtering, followed by a planarization process similar to that used in the production of

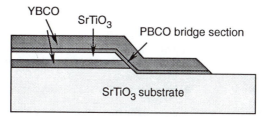

Fig. 13.11 Schematic diagram of a thin-film Josephson junction using ceramic superconducting YBCO.

silicon devices (see section 11.4.3). Currently, devices such as SQUIDs, gradio-meters and oscillators can be fabricated with junction areas as low as 0.7 μm × 0.7 μm. In the future it is possible that superconducting field-effect transistors (SUFETs) can be produced.

Light interacts with materials via the electrons. We have already considered this interaction in semiconductors and normal metals. The electrons in super-conductors are sensitive to light with photon energies as low as 0.01 eV and so superconducting materials can be used as photodetectors. These applications have been discussed by Doss [10] and by Richards and Hu [19].

The developments of devices described in this section represent the leading edge of low-current superconducting device technology at present.

REFERENCES

1. Kamerlingh Onnes, H. (1911) *Akad. van Wertenschappen* (Amsterdam), **14**, 113.
2. Gorter, C. J. and Casimir, H. B. G. (1934) *Physica*, **1**, 305.
3. London, F. and London, H. (1935) *Proc. Roy. Soc. Lond.*, **A149**, 71.
4. Bardeen, J., Cooper, L. N. and Schrieffer, J. R. (1957) *Phys. Rev.*, **108**, 1175.
5. Meissner, W. and Ochsenfeld, R. (1933) *Naturwiss.*, **21**, 787.
6. Bednorz, J. G. and Muller, A. (1986) *Z. Physik*, **B64**, 189.
7. Raveau, B. (1992) *Physics Today*, **45** (10), 53.
8. Ruggiero, S. T. and Rudman, D. A. (eds) (1990) *Superconducting Devices*, Academic Press, New York.
9. Orlando, T. P. and Delin, K. A. (1991) *Foundations of Applied Superconductivity*, Addison Wesley, Reading.
10. Doss, J. D. (1989) *Engineer's Guide to High Temperature Superconductivity*, John Wiley & Sons, New York.
11. Giaever, I. (1960) *Phys. Rev. Letts.*, **5**, 147.
12. Josephson, B. D. (1962) *Physics Letts.*, **1**, 251.
13. Anderson, P. W. and Rowell, J. M. (1963) *Phys. Rev. Letts.*, **10**, 230.
14. Clark, J. (1990) SQUIDS: Principles, Noise and Applications in *Superconducting Devices*, (eds S. T. Ruggiero and D. A. Rudman), Academic Press, New York.
15. Fink, H. J., Lopez, A. and Maynard, R. (1982) *Phys. Rev.*, **B26**, 5237.
16. Moschalkov, V. V., Gielen, L., Dhalle, M. *et al.* (1993) *Nature*, **361**, 617.
17. Van Duzer, T. and Turner, C. W. (1981) *Principles of Superconductive Devices and Circuits*, Elsevier/North Holland, New York.
18. Ralston, R. W., Kastner, M. A., Gallagher W. J. and Batlogg, B. (1992) *IEEE Spectrum*, **29** (8), 50.
19. Richards, P. L. and Hu, Q. (1989) *Proc. IEEE*, **77**, 1233.

FURTHER READING

Fitzgerald, K. (1988) Superconductivity: fact versus fancy, *IEEE Spectrum*, **25** (5), 30.
Orlando, T. P. and Delin, K. A. (1991) *Foundations of Applied Superconductivity*, Addison Wesley, Reading.
Ruggiero, S. T. and Rudman, D. A. (eds) (1990) *Superconducting Devices*, Academic Press, New York.
Van Duzer, T. and Taylor, C. E. (eds) (1989) *Proceedings of IEEE*, **77**, 1107.
Washburn, S. (1993) Nature, **361**, 587.
Advances in Applied Superconductivity: Goals and Impacts, a Preliminary Evaluation, Argonne National Laboratory, US Department of Energy, 1987.

14

Magnetic Materials – Magnetic Recording Technology

This chapter is concerned with a range of magnetic materials which are used to record both analog and digital information. These materials are now part of a large magnetic recording technology market which is expanding rapidly due to the demand for storage of digital information for use with computers. Consequently magnetic recording materials form the fastest-growing area in the magnetic materials market today. The objective of this chapter is to discuss the principles of magnetic recording and to show how this determines the properties of magnetic materials that can be usefully employed in magnetic recording. Some of the materials currently in use are discussed including those in widespread industrial use and some that are currently being developed or researched. The merits of the so-called 'perpendicular' recording media are considered relative to conventional recording media, and lastly the prospects for magneto-optic recording systems are assessed.

14.1 MAGNETIC RECORDING OF INFORMATION

What do we mean by magnetic recording?

Magnetic recording is the storage of analog or digital signals on a magnetic medium for subsequent retrieval and use. Magnetic recording exploits a very specific electronic property of certain materials. It uses the changes in orientations of magnetizations arising from large numbers of electrons with their spins aligned in such a way that small volumes of the material known as magnetic domains have a net spontaneous magnetic moment.

Magnetic recording technology spans a range of applications including relatively low-frequency analog audio recording, high-speed digital recording of data, and the recording of video information which uses the highest density of recording, 2660 flux reversals per millimetre [1]. Research has shown however that 5000 flux reversals per millimetre are possible [2], although such high densities are not yet commercially available.

Semi-permanent, that is non-volatile, storage of data for computers is achieved exclusively by magnetic methods today. The main types of recording media are magnetic tapes, floppy disks, hard disks and magneto-optic disks. Previously both magnetic bubble memory and magnetic drums were used for recording of data but these are no longer being employed in new computer systems.

Over the last fifty years, a large industry has built up based on magnetic recording of digital information (data) and analog signals (for example audio and video recording). The world production of materials for magnetic recording in 1982 amounted to \$235 million, which comprised \$31 million of γ-Fe_2O_3, tape, \$53 million of acicular γ-Fe_2O_3 for audio recording, \$108 million of cobalt-modified γ-Fe_2O_3 for video recording and \$43 million of CrO_2 mainly for use in video tapes. To this may be added \$50 million of soft ferrites for magnetic recording heads [3]. A growth rate of 12% per year [4] gives an estimated market of \$730 million per annum in 1992.

Therefore, although the market for magnetic recording media material is not yet as large as the markets for soft magnetic materials (1982 – \$1128 million, 1992 – estimated at \$3384 million) or for permanent magnets (1982 – \$680 million, 1992 – estimated at \$2040 million), it is still substantial and is actually growing more rapidly than the other two principal areas of magnetic materials because of the increasing use of computers and the need to store large amounts of data. The market for magnetic recording systems is however substantially greater than the market for the materials alone, and is currently about \$52 billion (10^9) per year [5].

14.1.1 Principles of magnetic recording

How is information stored in magnetic materials?

The arrangement for storing information on magnetic tapes is shown in Fig. 14.1. The principles of magnetic recording on magnetic tapes, floppy disks, hard disks and formerly on magnetic drums are the same. These are the principles of magnetic recording technology that we are concerned with here. The bubble devices were based on different principles, but this method of storage is now almost obsolete.

As a direct result of hysteresis in the magnetization curve of ferro- and ferrimagnets (Fig. 14.2), the final state of magnetization in zero field is dependent upon the magnetic history of the material. The remanent magnetization, or remanence, is a kind of memory of the last magnetic field maximum both in magnitude and direction. Therefore data, either in digital form for computers, or analog form as in audio recording, can be stored as a magnetic 'imprint' on a magnetic medium.

14.1.2 Criteria for magnetic recording media

What magnetic properties should the recording medium have?

The recording medium must have a relatively high coercivity so that the information stored is not accidently erased if the medium is unexpectedly

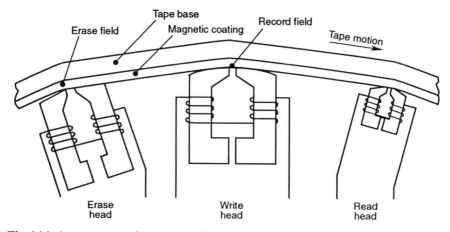

Fig. 14.1 Arrangement of 'erase', 'read' and 'write' heads in magentic tape recording.

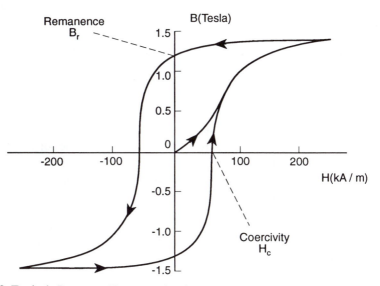

Fig. 14.2 Typical **B** versus **H** curve of a ferro- or ferrimagnet as used in magnetic recording media.

exposed to small magnetic fields. However, since the medium is likely to be re-used many times the coercivity must be small enough to allow erasure of information and subsequent re-use of the medium when desired. The medium must also have a relatively high saturation magnetization and remanence so that the leakage field from the surface of the material is large enough to be measured easily by the read head.

Coercivities of magnetic recording media are in the range of $20–150\,\mathrm{kA\,m^{-1}}$ $(250–1875\,\mathrm{Oe})$, while typical saturation magnetizations are in the range $0.3–2.0\,\mathrm{MA\,m^{-1}}$ $(300–2000\,\mathrm{emu\,cm^{-3}})$.

Table 14.1   Occurrence of wear problems in tapes and disks

Medium	*Head/medium contact*	*Many repetitions over same area*	*Wear problems*
Tapes	Yes	No	No
Floppy disks	Yes	Yes	Yes
Hard disks	No	Yes	No

Another problem that can arise in magnetic recording media is wear of the magnetic medium. This is particularly prevalent in floppy disks, as indicated in Table 14.1. The solution to this problem has been to incorporate solid lubricants into the fabrication of the floppy disks.

14.2   MAGNETIC RECORDING MATERIALS

Which are the most widely used materials for magnetic recording?

The materials used in different types of magnetic recording media [6] are described next.

Gamma iron oxide γ-Fe_2O_3

This is still the most popular magnetic recording material. It has been used in recording tapes since 1937. Now it is usually employed in the form of acicular particles of dimensions typically $0.1\,\mu m \times 0.5\,\mu m$ imbedded in a plastic binder. The material remains very cheap being a derivative of rust, although it is not a commonly occurring allotrope. Its coercivity ranges from 20–$30\,kA\,m^{-1}$ (250–$375\,Oe$) and its saturation magnetization is $370\,kA\,m^{-1}$. The Curie temperature of $600\,°C$ is well above ambient temperature which therefore avoids loss of recording information due to temperature variations except in very extreme cases.

Chromium dioxide CrO_2

This was formerly popular as a high-performance magnetic recording material, but it is no longer widely used. It has higher coercivity than gamma ferric oxide, but a lower Curie temperature. It was used in the form of acicular particles of dimensions $0.05\,\mu m \times 0.4\,\mu m$. This had the advantage of giving higher recording densities and better signal-to-noise ratio because of its higher saturation magnetization and coercivity. Typical coercivity values are 40–$80\,kA\,m^{-1}$, with a saturation magnetization of $500\,kA\,m^{-1}$ and a Curie temperature of $128\,°C$. The low Curie temperature caused some problems with thermal stability of magnetization and consequent loss of recorded information.

Cobalt surface-modified γ-Fe$_2$O$_3$

This material is used increasingly today for magnetic recording media. It consists of a cobalt layer 3 nm thick on the surface of gamma iron oxide particles. This leads to increased coercivity. The typical value of coercivity of this material is 50 kA m^{-1} and the saturation magnetization is typically 370 kA m^{-1}.

Hexagonal ferrites

These are usually fabricated from barium or strontium ferrite. They have high coercivity and are used for 'permanent' recording of data for credit card applications. In these cases erasure is seldom required and should be avoided as far as possible.

Ferromagnetic powders

These recording materials have high saturation with M_s up to 1700 kA m^{-1} in iron powders, giving good signal-to-noise ratios. The coercivities are typically 120 kA m^{-1}.

Metallic films

These are widely used in hard disks because they can be relatively easily deposited on the surface of the rigid disk and they have high saturation and remanence. Coercivities range from 60–100 kA m^{-1} and saturation magnetizations can be up to 1 MA m^{-1}.

Perpendicular media

Much research is being conducted into perpendicular recording media at present. Materials such as barium ferrite and cobalt–chromium films are under investigation as possible perpendicular recording media. The materials still suffer from problems, which are described in section 14.3.4. Consequently there is little development of these materials in commercial applications yet.

14.2.1 Saturation magnetization of γ-Fe$_2$O$_3$

How can we calculate the saturation magnetization of a ferrimagnetic recording material such as gamma iron oxide from the known unit cell of the material?

We know from earlier discussion that the magnetic moment of a magnetic material is dependent on unpaired electron spins. In the case of magnetic recording media, these electron spins are localized at the ionic sites in the material. The most common material used in magnetic recording is γ-Fe$_2$O$_3$

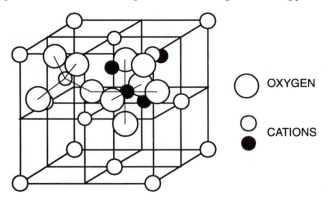

OXYGEN

CATIONS

Fig. 14.3 Unit cell of gamma iron oxide, the most widely used magnetic recording material.

(gamma iron oxide). This is a ferrimagnet which means that not all magnetic moments in the unit cell are aligned parallel.

Gamma iron oxide has $2.5\,\mu_B$ per Fe_2O_3 molecule. The unit cell has a spinal structure and contains 32 oxygen atoms as shown in Fig. 14.3. The oxygen atoms form a cubic lattice. The Fe^{3+} ions can occupy two types of sites, A and B. The A sites are surrounded by four oxygen atoms and B sites are surrounded by six oxygen atoms. There are 8 occupied A sites and 16 possible B sites in a unit cell. Of these B sites, on average 13.33 ($= 40/3$ exactly) out of the sixteen are occupied.

The exchange interaction in this material aligns the magnetic moments of the iron atoms on the A sites and B sites antiparallel. Since the magnetic moment on Fe^{3+} is $5\,\mu_B$ the net magnetic moment of a unit cell is,

$$m = (\tfrac{40}{3} - 8)5\mu_B$$

$$= \tfrac{80}{3}\mu_B$$

and since there are (32/3) molecular units of Fe_2O_3 per unit cell, this gives a net magnetic moment of $2.5\,\mu_B$ per molecule of Fe_2O_3. This is about one quarter of the value which would be expected if the Fe^{3+} ions were all aligned with magnetic moments parallel (i.e. ferromagnetic) in $\gamma\text{-}Fe_2O_3$.

14.3 CONVENTIONAL MAGNETIC RECORDING USING PARTICULATE MEDIA

How are the magnetic particles arranged in the recording medium?

In conventional magnetic recording, whether digital or analog, the medium is usually one of the seven different types of material listed in section 14.2. The particulate media, which are used most often, have single-domain particles

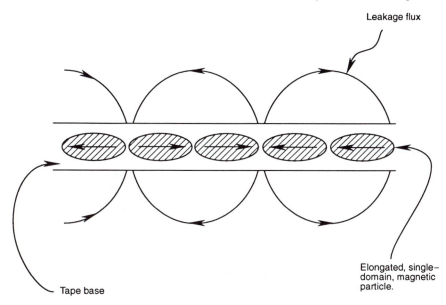

Leakage flux

Elongated, single-
domain, magnetic
particle.

Tape base

Fig. 14.4 Arrangement of acicular magnetic particles in conventional 'longitudinal' magnetic recording media. (Schematic only – particle sizes are not drawn to scale.)

embedded parallel to the surface of the medium as shown in the above diagram (Fig. 14.4).

The medium is magnetized using a small electromagnet called the recording head or 'write' head. This is usually made of a soft ferrite material or permalloy and has a small pole gap of typically 0.8 µm. The recording head generates a longitudinal field in the medium which is sufficient to change the direction of magnetization of the particles in the tape. Media in which the particles are aligned parallel to the surface are known as longitudinal media. Media with the alternative configuration, with the particles aligned normal to the surface, are known as perpendicular recording media. These are discussed in section 14.3.4.

14.3.1   The read–write head

How is a field impressed on the magnetic recording medium and how is the magnetization of the medium detected?

The read–write head is a high-permeability magnetic circuit usually fabricated from soft ferrite or permalloy, with a narrow gap in the circuit which can be used to generate the fringing field for writing, or to collect the flux in the reading process. The width of the gap (typically 0.8 µm) determines the recording density, along with other factors, including the characteristics of the recording medium.

The field in the head gap is one of the most critical factors in magnetic recording. Much effort has been, and continues to be, devoted to calculating the fields in the gap in order to optimize performance of magnetic recording systems. The simplest example involved the use of a highly-idealized read head, known as the Karlqvist head [7] for which calculations could be made under simplified conditions.

14.3.2   The writing (or recording) process

How is information 'written' onto the medium?

We shall discuss in general terms how the medium is magnetized. A current is passed through the write-head coil which generates a field in the head gap. In a tape recording system the magnetic tape passes the recording head gap, where it is subjected to the fringing field. This leads to a change in the state of magnetization of the tape depending on the strength and direction of the magnetic field in the head gap. The magnetic imprint on the tape, in the form of remanent magnetization, is therefore a record of the magnetic field in the gap when the tape passed it. As a simplified example the remanence in one direction B_{R+} could represent the state '1' and remanence in the other direction B_{R-} could represent the state '0' in digital magnetic recording.

So far it sounds fairly simple. In fact the writing process itself is not well understood, even though it works very well in practice. The main cause of difficulty is the exact relation between the signal recorded on the tape, that is the magnetization, and the field in the head gap at the time the tape passed. The recording process is non-linear and hysteretic and therefore extremely difficult to model in a satisfactory way.

The magnetization curve, or hysteresis loop, of these materials is known to be difficult to model accurately. Furthermore, as the tape passes the head it experiences differences in field strength and direction depending not only on its distance from the gap, but also on the changes in the gap field with time as the signal changes. Therefore with these complications our understanding of the process is very poor and it is difficult to predict the final magnetization of the medium.

The recoil minor loops which the medium traverses as any given region of the tape moves away from the head gap are not simple recoil loops because the field in the gap is also changing with time. Therefore the field experienced by such a region changes in a complicated way with both time and position.

In order to improve the linearity of the remanent magnetization on the tape with gap field, the writing process is usually made under AC bias conditions. That is to say the signal level which we wish to record on the tape is supplied to the recording head in the form of a current which is proportional to the signal (the DC bias) and on the top of this an AC signal is superimposed. The resulting magnetization curve, shown in Fig. 14.5, is known as the anhysteretic remanent magnetization curve (ARM). This overcomes many of the inherent

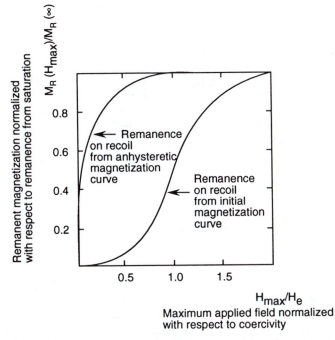

Fig. 14.5 Anhysteretic remanent magnetization curve used for 'linearizing' the response of the magnetization of the medium to the external field [1]. Reproduced with permission of R. Bertram, *Proc. IEEE*, **74**, 1986, p. 74.

non-linearities of the magnetization curve. The procedure is universally used in analog sound recording.

This linearization of the relation between magnetization and field is highly desirable, particularly in analog recording processes such as audio recording. It means that the signals generated in the read or 'playback' mode will be proportional to the original input signal.

14.3.3 The reading process

How is information 'read' from the magnetic medium?

The reading process in magnetic recording consists of the movement of a tape or disk past the 'read head' of the recorder. The magnetic signals encoded on the tape lead to changes in leakage flux from the surface of the tape. This flux passes through the body of the read head and in turn can be detected by a coil wrapped around the head as shown in Fig. 14.6. The read head must be magnetically very soft with high permeability so that the external field can give a large change in B the flux density in the head and hence a large voltage in the pick-up coil. The voltage that is generated in the coil is governed by the Faraday–Lenz law of electromagnetic induction.

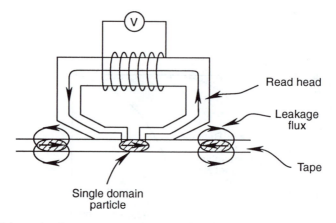

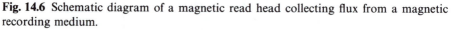

Fig. 14.6 Schematic diagram of a magnetic read head collecting flux from a magnetic recording medium.

The leakage flux from the magnetic tape can be found from the magnetization geometry of the particles in the tape. This flux is then captured by the read head and the voltage in the coil is proportional to $d\Phi/dt$ in the head. This means that the voltage output from the head can be related to the magnetization on the tape. Note that in this case we do not need to know anything about the magnetization mechanism in the tape. This makes the reading process much easier to analyse than the writing process.

14.3.4 Perpendicular recording: prognosis

What are the prospects for successful implementation of perpendicular recording in operational systems?

Perpendicular recording, in which the single-domain particles are aligned normal to the plane of the tape should have several advantages over the present longitudinal recording as discussed for example by Hubner *et al.* [8]. The main advantages should be:

- higher storage density (bits per cm)
- higher and more sharply-defined leakage field.

The configurations of magnetic particles in a perpendicular recording medium are shown in Fig. 14.7. Two types of material have been under investigation for use in perpendicular recording media:

- sputtered cobalt-chromium film (Co–18% Cr)
- barium ferrite.

These media have suffered from problems which have so far rendered them

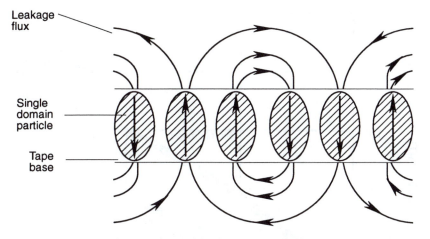

Leakage flux

Single domain particle

Tape base

Fig. 14.7 Arrangement of magnetic particles in a 'perpendicular' (or 'vertical') recording medium. (Schematic only – particle sizes are not drawn to scale.)

unworkable except on a research basis. These problems have ranged from difficulties with the mechanical properties and stability of the medium, poor signal-to-noise ratio arising from excessively high noise on playback, and the need for a very small head-to-medium distance ('flying height') which is less than present technology can handle. Research on this type of magnetic recording technology continues mainly in Japan [9]. Recent work on the use of acicular barium ferrite particles for perpendicular recording media has been reported by Sueto *et al.* [10].

14.3.5   Storage densities

How rapidly are storage densities increasing and which recording methods offer the greatest prospects for further increases?

As may be expected, it is advantageous to increase the storage densities of disks and tapes [11]. This is particularly important for computer disk drive systems since it means that more data can be stored in a given disk space. The storage density in a medium depends on the magnetic properties of the medium and the characteristics of the write head. The linear bit density on the track of a magnetic disk is determined by the transition length 'λ' which is the minimum distance that is needed to completely reverse the magnetization from saturation remanence in one direction to saturation remanence in the other direction.

On magnetic disks the linear bit density is still usually measured by convention in bits per inch. The number of bits per inch times the number of tracks per inch gives the storage density in bits per square inch. By 1980, the storage density on flexible disks had reached 8×10^5 bits/sq. in. (128×10^3 bits/cm^2),

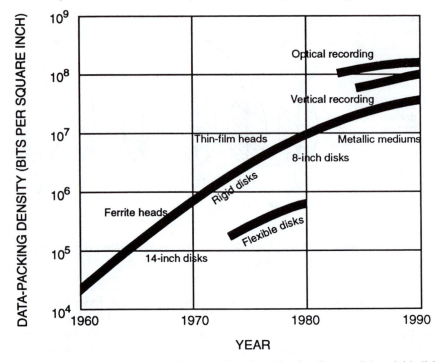

Fig. 14.8 Improvements in magnetic recording densities for floppy disks, rigid disks, magneto-optical and perpendicular (vertical) recording techniques.

while on hard disks the density was 1×10^7 bits/sq. in. (1.6×10^6 bits/cm^2). By 1990 these storage densities had improved by a factor of 5.

Perpendicular recording has an expected density of 10^8 bits/sq. in. (16×10^6 bits/cm^2) which is comparable with the storage densities expected from magneto-optic recording. Work by Nakamura *et al.* [12] has shown that on 'perpendicular' hard disk systems densities of over 10^6 bits mm^{-2} ($= 6.25 \times 10^8$ bits/sq. in) are possible. However, continual improvements in conventional longitudinal recording have been made, so that it is questionable whether 'perpendicular' recording systems have any intrinsic advantages over conventional recording technology.

The use of magnetoresistive read heads for detecting the signals imprinted on the medium seems to offer the best prospects for future high-performance, high-density magnetic recording systems, as discussed by Edelman *et al.* [13].

14.4 MAGNETO-OPTIC RECORDING

What are the main advantages of magneto-optic recording systems over conventional recording systems?

The high storage densities achievable through magneto-optic recording make it an attractive option for storage of digital data on computer disk drive systems.

In 1983 Philips had developed an early 10 megabyte magneto-optic disk with a packing density of 4×10^6 bits cm^{-2} (25×10^6 bits/sq. in) which was twice that obtained on rigid disks. More recent developments suggest that magneto-optic disks for computer systems having storage capacities of perhaps 30 gigabytes and data transfer rates of 8 megabytes per second will be available on commercial personal computers in the next 5–10 years [14].

14.4.1 Mechanism of magneto-optic recording

How does magneto-optic recording work?

Data is written on a magneto-optic disk using a laser beam. The disk consists of a magnetic material with relatively low ordering temperature and a relatively high angle of Kerr rotation. This material is heated by incident laser light which

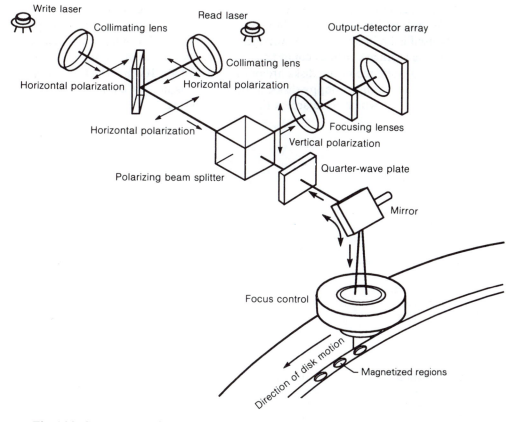

Fig. 14.9 Components of a magneto-optic disk recording system. Reproduced with permission of R. M. White, *Introduction to Magnetic Recording*, published by IEEE Press, 1985.

raises the temperature and reduces the coercivity. The material is then exposed to a magnetic field which can alter the direction of magnetization of the region, as a result of the reduced coercivity. The region is then allowed to cool as it passes out of the laser beam and the coercivity rises leading to semi-permanent storage of data. Recent developments have revealed new materials with extremely large Kerr angles which look to have high potential for future applications in magneto-optic recording systems [15, 16].

The reading process must take place without corrupting the data. This is achieved by use of the magneto-optic Kerr effect in which a linearly polarized beam of laser light has its direction of polarization rotated, as a result of reflection, depending on the direction of magnetization in the reflecting medium. The size of the angle of rotation depends on the material used. This allows the direction of magnetization at various points on the magneto-optic disk to be determined. The laser beam that is used for reading has a lower intensity than that used for writing, in order to ensure that the data is not corrupted by overheating the region where the data is stored.

The magneto-optic method of data storage has better signal-to-noise ratios than conventional magnetic recording for the same storage density. The storage density itself is limited only by the wavelength of light used. If ultra-violet lasers are used the storage densities could be as high as $500\,\mathrm{Mbits\,cm^{-2}}$. Access times for magneto-optic recording disks are of the order of 40–100 ms which is about ten times faster than for floppy disks, although this is not yet competitive with Winchester disks which have access times of 20–60 ms.

Table 14.2 Comparison of speeds and recording densities of different magnetic recording techniques [1]

Medium density systems	Speed $(m\,s^{-1})$	Max f (kHz)	Min λ (μm)	Signal
Professional audio	0.38	20	18.75	Linear
Computer tapes	1.14	36	31.8	Digital
Computer disks	25	4500	5.5	Digital
High density systems				
Instrumentation	0.05–2.0	2000	1.5	Digital
Quad videos	38	15000	2.5	FM
Consumer videos	5.5	7000	0.75	FM
Audio cassettes	0.05	20	2.0	Linear
Future computer disks	25	10000	2.5	Digital

14.4.2 Comparison of speeds and densities of various recording methods

How fast can data be accessed by the different types of recording systems?

The several types of magnetic recording technique have different speeds, maximum operating frequencies and storage densities. These have been compared by Bertram [1] and are shown in Table 14.2.

REFERENCES

1. Bertram, H. N. (1986) *Proc. IEEE*, **74**, 1494.
2. Tsang, C., Chen, M. M., Yogi, T. and Ju, K. (1990) *IEEE Trans. Mag.*, **26**, 1689.
3. White, R. M. (1985) *J. Appl. Phys.*, **57**, 2996.
4. Psaras, P. A. and Langford, H. D. (1987) *Advancing Materials Research*, National Academy Press, Washington DC, p. 91.
5. Hoyt, R. F. (1992) *IEEE Mag. Soc. Newsl.*, **29**, 1.
6. Jiles, D. C. (1991) *Introduction to Magnetism and Magnetic Materials*, Chapman & Hall, London, Chapter 14, p. 327.
7. Karlqvist, O. (1954) *Trans. Roy. Inst. Tech. (Stockholm)*, **86**.
8. Hubner, R., Schewe, H., Zintl, W. and Rockelein, R. (1992) *J. Magn. Magn. Mater.* **104**, 965.
9. Ouchi, K. and Iwasaki, S. (1987) *IEEE. Trans. Mag.*, **23**, 180.
10. Sueto, K., Sakamoto, H., Suzuki, A. and Sugimoto, M. (1992) *J. Magn. Magn. Mater.*, **104**, 979.
11. White, R. M. (1983) *IEEE Spectrum*, **20**(8), 32.
12. Nakamura, Y. and Muraoka, H. (1991) *IEEE Trans. Mag.* **27**, 4555.
13. Edelman, H., Brock, G. W., Carr, T. *et al.* (1990) *IEEE Trans. Mag.*, **26**, 3004.
14. Yamada, Y. *et al.* (1991) *IEEE Trans. Mag.*, **27**, 5121.
15. Di, G. Q., Iwata, S., Tsunashima, S. and Uchiyama, S. (1992) *J. Magn. Magn. Mater.*, **104**, 1023.
16. Qingqi, Z., Zhi, Z., Wuyan, L. *et al.* (1992) *J. Magn. Magn. Mater.*, **104**, 1019.

FURTHER READING

Gambino, R. J. and Suzuki, T. *Magneto-Optical Recording Materials*, to be published by IEEE Press, New York.

Kryder, M. H. *Magneto-Optic Recording Fundamentals* to be published by IEEE Press, New York.

Mallinson, J. C. (1987) *The Foundations of Magnetic Recording*, Academic Press, San Diego.

White, R. M. (1984) *Introduction to Magnetic Recording*, IEEE Press, New York.

15

Electronic Materials for Transducers – Sensors and Actuators

In this chapter we will be concerned mainly with materials that are used in energy conversion devices. By far the most common are those which convert electrical into mechanical energy and vice versa. These are the electrostrictive/piezoelectric transducers. Most of the discussion of electrostrictive/piezoelectric transducers can also be applied to magnetostrictive transducers, which are mentioned where they can be treated analogously to the electrostrictive/piezoelectric materials. We will look at the different types of materials for these transducers and define their performance parameters. Among these materials the most important class is the ferroelectrics which have high relative permittivities and high strain coefficients.

15.1 TRANSDUCERS

What do we mean by the term transducer?

A transducer is any device that converts one form of input energy into a different form of output energy. In other words it is an energy conversion device. A common example is a device which converts mechanical energy into electrical energy, such as a piezoelectric transducer. In fact the original use of the term transducer was specifically for a device which sensed mechanical input energy and converted it into electrical output energy. However the term is now used to include any device which converts one form of energy to another. The efficiency of a transducer is a useful parameter which measures the ratio of output energy to input energy. Examples of transducers are loudspeakers, ultrasonic vibration generators, thermocouples, microphones and various forms of magnetometer.

Except for some rather minor differences of meaning the terms transducer, sensor and detector are regarded as synonymous. In these transducers the output bears a known relation to the input. Therefore the output can be controlled through the input or alternatively the input determined by measuring the output. Another way of viewing transducers is simply as a means of interacting between electronic instrumentation and the outside world. In most cases therefore we are concerned with devices that act as detectors (sensors) which convert the external energy into an electrical voltage, or emitters (actuators) which convert the electrical voltage into external energy.

15.1.1 Classification of transducers

What terminology is used for the various types of transducers?

The general categories of transducers are shown in Table 15.1 below.

The non-linearity of a transducer can be an important factor. Non-linearity refers to the deviation from a directly proportional dependence of output on input, for example when the mechanical force produced by an electrostrictive transducer varies with the square of the applied electric field. Ferroelectric and ferromagnetic materials (sections 15.4 and 14.1 respectively) also exhibit a hysteretic relationship between input and output which can be a hindrance in certain transducer applications, particularly for positioning devices.

The most common form of transducers are those which convert to and from acoustic energy. These are used in loudspeakers, microphones and acoustic systems. The three types of acoustic transducer that are widely used are piezoelectric, electrostrictive and magnetostrictive. These acoustic transducers are usually operated at, or close to, mechanical resonance in order to obtain maximum energy conversion efficiency. For wide bandwidth operation the transducer must necessarily be operated at frequencies well away from the resonant frequency. This is achieved in some cases by sandwiching layers of transducer material with plates of metal to reduce the sharpness of the resonance.

There is a wide range of physical sizes for transducers from a few square millimetres and weights of a few grams up to square metres and weights of tens of kilograms. Transducer-phased arrays are sometimes used instead of a single transducer in order to steer the acoustic beam in particular directions. In some

Table 15.1 Nomenclature of transducers in terms of their forms of input and output energy

Type of transducer	Input energy	Output energy
Electrostrictive	Electric	Mechanical
Piezoelectric	Mechanical	Electrical
Electroacoustic	Electrical	Sound
Photoelectric	Light	Electrical
Magnetoelectric	Magnetic	Electrical
Thermoelectric	Thermal	Electrical
Magnetostrictive	Magnetic	Mechanical
Piezomagnetic	Mechanical	Magnetic
Pyroelectric	Infra-red radiation	Electrical

cases special transducer arrays several metres across and weighing about a ton have been constructed for submarine surveillance applications.

15.1.2   Energy conversion in transducers

What properties must a material have to make it suitable for use as a transducer?

In order for a material to be useful as a transducer it must have a high efficiency for converting one form of energy into another. The input energy is usually supplied by an external influence such as an applied stress, an electric field or a magnetic field. The response of the material, in the form of a voltage or a strain, can be measured and used to detect the external field, as in a sensor, or to cause movement, as in an actuator.

The most common forms of transducer material are piezoelectric, electrostrictive and magnetostrictive materials. Piezoelectric/electrostrictive transducer materials include quartz, ammonium dihydrogen phosphate, tourmaline, lithium sulphate, barium titanate, lead zirconate and lead magnesium niobate, and Rochelle salt (potassium sodium tartrate tetrahydrate, $KNaC_4H_4O_6.4H_2O$). Magnetostrictive transducer materials include nickel alloys and some of the more recent rare earth–iron alloys such as Terfenol (terbium-dysprosium-iron).

15.2   TRANSDUCER PERFORMANCE PARAMETERS

What material properties are important for transducer applications?

Ferroelectrics are widely used as a transducer material, since they give a high polarization-to-field-strength ratio. These materials need to be polarized or 'poled' in order to ensure that the piezoelectric effect is observed. On the other hand the dependence of strain on electric field is usually relatively high in these materials, although it is also hysteretic as shown in Fig. 15.1. This is analogous to the variation of bulk magnetostriction with magnetic field, and produces the well-known 'butterfly curve' of strain versus field.

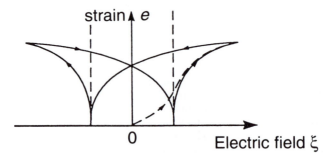

Fig. 15.1 Dependence of strain on electric field in a ferroelectric material showing hysteresis. Reproduced with permission of N. Braithwaite and G. Weaver, *Electronic Materials*, published by Butterworths, 1990.

15.2.1   Strain

How do we define electrostrictive strain in these materials?

The strain is given simply as the fractional change in length,

$$e = \Delta l / l_0$$

where l_0 is the original length and Δl is the change in length. The length l_0 is usually taken for convenience as the length in the depolarized state (or in magnetostrictive materials the demagnetized state) and the changes are therefore measured by convention relative to the depolarized (or demagnetized) state. The above equation is itself merely a definition and tells us nothing about the performance of the materials or how the strain varies with field or polarization. The strain derivatives, $de/d\xi$ for electrostrictive/piezoelectric materials and de/dH for magnetostrictive materials, are rather more meaningful representations of material properties, since they tell us how rapidly the strain changes with the relevant applied field.

15.2.2   Motor coefficient

What does the derivative of strain with respect to field tell us about the suitability of a material for transducer applications?

The derivative of the strain with respect to the applied field at constant stress is one of the most useful parameters for quantifying the performance of a transducer material. This is called the motor coefficient d, which for piezoelectric/electrostrictive materials is given by

$$d = \left(\frac{\partial e}{\partial \xi} \right)_\sigma$$

and is measured in $V^{-1} m$ (or equivalently $C N^{-1}$). For magnetostrictive materials it is given by

$$d = \left(\frac{\partial e}{\partial H} \right)_\sigma$$

and is measured in $A^{-1} m$.

When changes expressed by the differentials are small and reversible, these d coefficients are also equal to $(\partial D/\partial \sigma)_\xi$ and $(\partial B/\partial \sigma)_H$ respectively.

This is also sometimes known as the strain coefficient, the strain derivative or sometimes simply as the 'd coefficient'. It is a useful measure of the performance of a material, since as we shall see, it is closely related to the energy transfer efficiency of the material. It is often quoted in the range of picometers per volt for piezoelectric/electrostrictive materials or nanometers per amp in magnetostrictive materials. Typical values of d for various materials are: lead zirconate titanate 5×10^{-10} V^{-1}m, ammonium dihydrogen phosphate 5×10^{-11} V^{-1}m, lead magnesium niobate 1.5×10^{-9} V^{-1}m, iron 3×10^{-9} A^{-1}m, terbium dysprosium iron 30×10^{-9} A^{-1}m.

At the origin of the hysteresis curve, that is in the unpolarized (unmagnetized) state, the strain derivative is very small, as can be seen by reference to Fig. 15.1. This condition is not very useful for a transducer. However the material can be used in its 'poled' (magnetized) condition whereupon the strain derivative will be considerably larger and the strain will remain almost linear with change in field for a range of field strengths. In active devices it is also possible to use the material under a field bias which can be adjusted to find the maximum of $de/d\xi$ or de/dH.

Most piezoelectric transducers are ferroelectrics which are 'poled' to produce a remanent polarization and then are operated over a range of electric fields which is small enough to ensure that depolarization does not occur. The exceptions are the single-crystal piezoelectrics which are not ferroelectric and therefore cannot be 'poled'.

15.2.3 Generator coefficient

How is the generation of field by applied stress quantified?

When using a piezoelectric material to generate a voltage by application of a stress, the parameter of interest is the rate of change of electric field ξ with stress. This is called the generator coefficient g, which is defined at constant electric flux density D,

$$g = -\left(\frac{\partial \xi}{\partial \sigma}\right)_D = \frac{1}{\varepsilon_0}\left(\frac{\partial P}{\partial \sigma}\right)_D$$

This is measured in VN^{-1}m (or equivalently in m^{2}C^{-1}). The analogous coefficient for a magnetostrictive transducer is,

$$g = -\left(\frac{\partial H}{\partial \sigma}\right)_B = \left(\frac{\partial M}{\partial \sigma}\right)_B$$

where H is the magnetic field and B is the magnetic flux density. This is measured in AN^{-1}m.

When the changes represented by the differentials are small and reversible, these g coefficients are also equal to $-(\partial e/\partial D)_\sigma$ and $-(\partial e/\partial B)_\sigma$ respectively.

This is sometimes called the piezoelectric coefficient or polarization coefficient. It is often quoted in the range of millivolt meters per Newton. Typical values of g for different materials are: lead zirconate titanate 30×10^{-3} VmN^{-1}, 'PXE' (a commercial modified form of lead zirconate titanate) 10×10^{-3} VmN^{-1}, potassium sodium niobate 20×10^{-3} VmN^{-1}.

15.2.4 Energy coupling coefficient

How can the energy conversion efficiency be represented?

One of the most useful parameters for a transducer, whether it is electrostrictive or magnetostrictive, is its energy conversion efficiency. This can be expressed via the square of the coupling coefficient, k^2 which is defined as

$$k^2 = \frac{\text{energy output}}{\text{energy input}}.$$

In the case of an electrostrictive transducer the value of k^2 is given by

$$k^2 = \frac{E_Y}{\varepsilon_0 \varepsilon_r} d^2$$

where d, the strain derivative, has been defined in section 15.2.2. In the case of a magnetostrictive transducer k^2 is given by

$$k^2 = \frac{E_Y}{\mu_0 \mu_r} d^2.$$

However it must be remembered that in both ferroelectric and ferromagnetic materials, ε_r and μ_r are not constants. This means that the energy conversion efficiency changes with applied field strength. Therefore it is not a material constant.

15.3 TRANSDUCER MATERIALS CONSIDERATIONS

What considerations determine the choice of a particular material for a transducer?

The first problem in fabricating a transducer for a particular application is to find a material which exhibits the right effect for converting the input energy into the required output energy. Once this has been achieved it is usually necessary to modify the material itself to optimize its performance. This means enhancing the desired properties to increase the energy conversion efficiency. This can be achieved in some cases by adjusting the chemical composition. For example in recent magnetostrictive transducer materials the anisotropy has been reduced by suitable choice of alloying components leading to an increase in the rate of change of strain with applied field.

15.3.1 Piezoelectricity

Can a voltage be induced in a material by the application of stress?

The piezoelectric effect is the production of an electric charge on the surface of a material, and hence a voltage across the material, as a result of the application

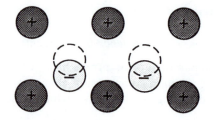

Fig. 15.2 Non-centrosymmetric crystal classes can exhibit piezoelectricity because the application of a stress changes the separation between positive and negative ions leading to a net polarization.

of stress. This was first discovered by J. and P. Curie [1]. The piezoelectric effect therefore allows a conversion of mechanical energy into electrical energy through a material transducer.

This occurs in materials where the application of stress causes a change in electric polarization by separating the centres of positive and negative charge in the crystal. In crystalline materials the piezoelectric effect only occurs in a limited class of materials of low crystal symmetry in which the application of stress deforms the crystal structure and leads to the generation of an electric dipole moment as shown in Fig. 15.2. These materials necessarily have a crystal structure which lacks a centre of symmetry [2, p. 273].

The converse effect also occurs: when an electric field is applied to a piezoelectric material a strain is produced. This is somewhat similar to electrostriction (section 15.3.2), but in this case the strain is antisymmetric with respect to the electric field, which means that it is proportional to odd powers of the field strength. Therefore the strain in a piezoelectric material changes sign when the electric field is reversed in direction. In the majority of cases piezoelectric materials exhibit greater strain under the action of an electric field than conventional electrostrictive materials, although some of the ferroelectric electrostrictive materials exhibit comparable strains to piezoelectric materials.

One of the most important applications of piezoelectricity is in the quartz crystal resonator, in which the strain amplitude can become very large when the applied AC voltage signal coincides with the mechanical resonance of the quartz crystal. Similarly when the frequency of mechanical excitation coincides with the resonant frequency, a large electrical signal is produced.

15.3.2 Electrostriction

What happens to the shape of a dielectric when it is subjected to an electric field?

All dielectric materials undergo a strain when subjected to an applied electric field, resulting in a slight change in shape. The change in length of a dielectric under the action of an electric field was first discovered by Lippmann [3]. This change in length is termed electrostriction.

The exact definition of electrostriction is somewhat more limited than simply the strain produced by an applied electric field. In order to distinguish it from being the converse of the piezoelectric effect, the term 'electrostriction' is sometimes used to refer only to those strains which do not change sign if the direction of the electric field is reversed.

Some materials, such as the ferroelectrics, can exhibit either piezoelectricity or electrostriction depending on the conditions under which they are operated. In fact electrostriction always occurs in dielectric materials under the action of an electric field, but when both types of strain occur together the electrostrictive strains are much smaller than the piezoelectric strains. In these cases we simply refer to the material as piezoelectric.

Compared with the antisymmetric strains induced in the inverse piezoelectric effect the symmetric electrostrictive strains are in most cases rather small, being typically $\lambda \approx 10^{-4}$ [3,4]. In some ferroelectrics however the electrostrictive strains can be much larger, for example $\lambda \approx 10^{-2}$–10^{-3} in lead zirconate titanate [5,6] or lead magnesium niobate [7]. This latter material, which at lower temperatures is a ferroelectric, is usually operated above its Curie temperature in transducers.

15.3.3 Magnetostriction

What happens to the shape of a magnetic material when it is subjected to a magnetic field?

Magnetostriction is a property of materials which change length when they are magnetized either spontaneously by virtue of a magnetic phase transition (spontaneous magnetostriction) or under the action of a magnetic field (field-induced magnetostriction). This effect is only significant in magnetically ordered materials such as ferromagnets, ferrimagnets and antiferromagnets. It was first discovered by Joule [8]. Typical magnetostrictive strains of magnetic materials are of the order of $\lambda \approx 10^{-5}$–10^{-6} [9], although more recent advanced materials, such as terbium-dysprosium-iron, have magnetostrictions as high as 10^{-3} [10,11]. If the ferromagnet is demagnetized the magnetostrictive strains are symmetric with applied field, but if the ferromagnet is in a remanent magnetized state the magnetostrictive strains are antisymmetric with applied field.

15.3.4 Piezomagnetism

Is there an inverse effect to magnetostriction?

Under certain conditions the application of stress to a magnetic material can cause a change in magnetization. In unmagnetized materials this is the magnetic analogue of the piezoelectric effect in unpolarized dielectric materials. Only a few examples are known of magnetic materials which can change from a demagnetized to a magnetized state under the action of a stress. One of these is the antiferromagnet CoF_2 in which a small magnetization of the order of $10^3\,A\,m^{-1}$ can be produced by large shear stresses [12].

A related, and much more common, phenomenon is the tendency of magnetic materials to change their magnetization under the action of an applied stress when already magnetized [13]. This effect is only of significant size in ferromagnets and is rarely employed in applications. There are no known cases of ferromagnets which become magnetized under the action of a stress on the unmagnetized state.

15.3.5   Mechanism of piezoelectricity

Why does a voltage change arise in a piezoelectric as a result of stress?

The phenomena of piezoelectricity and electrostriction are of great practical importance in electromechanical transducers. Consider a piezoelectric material which elongates along the direction of polarization. If this material is subjected to a compressive stress along the axis of polarization, and if we consider only reversible processes, the material can best respond to the compressive stress by polarizing itself at right angles. This will minimize the strain energy of the system and leads to ionic displacements at right angles to the original direction of polarization and compressive stress. The result is the generation of a voltage along the direction perpendicular to the stress where there was previously no voltage. Therefore the compressive stress has generated a change in voltage. This is the piezoelectric effect.

The inverse effect is the change in strain along the axis of polarization caused by the application of an electric field. Suppose an electric field is applied perpendicularly to the direction of polarization. This will eventually result in a rotation of polarization into the field direction, and this causes a change in strain along the direction of the field. By the Poisson effect there will also be a change in strain at right angles to this direction.

15.3.6   Comparison of electrostriction and magnetostriction

Are electrostriction and magnetostriction analogous effects under the action of electric and magnetic fields respectively?

According to the conventional definition, electrostrictive strain is dependent on even powers of the electric field strength (i.e. symmetric with field strength). This definition is suitable for most dielectrics, but can cause a problem in ferroelectrics (section 15.4), because of hysteresis in the relationship between polarization and electric field. If we wish to maintain the analogy with magnetostriction, the electrostriction should be defined as a strain which is invariant under the reversal of the direction of polarization, because the magnetostrictive strain can be either symmetric or antisymmetric with field strength, depending on whether the magnetic material was already magnetized.

The analogy with magnetostriction is clear. In many cases ferromagnetic magnetostrictive transducers and actuators are operated in a 'biased' condition which is equivalent to the 'poled' condition of a ferroelectric electrostrictive transducer or actuator. The objective of biasing (poling) is merely to find a point on the magnetostriction (electrostriction) curve at which the strain derivative is greater than at the unmagnetized (unpolarized) state. The dependence of strain on magnetization (polarization) remains symmetric despite the biasing. However the dependence of strain on magnetic field (electric field) becomes antisymmetric once the material has been biased (poled). In magnetic materials the phenomenon is referred to as magnetostriction under both conditions as there has not been any fundamental change in the nature of the material or its response to the applied field.

15.3.7   Piezoelectric and piezomagnetic response times

What is the relative speed of response in piezoelectric/electrostrictive and magnetostrictive materials?

Generally the speed of the piezoelectric response is much faster than the analogous response of a magnetic material to field or stress. Piezoelectric transducers can operate at frequencies of up to tens of megahertz whereas magnetostrictive transducers are restricted to a few kilohertz. This is because most highly electrostrictive materials are electrical insulators, while most highly magnetostrictive materials are electrical conductors. In this latter case the generation of eddy currents at higher frequencies limits the penetration of the alternating magnetic field into the material.

Fig. 15.3 Schematic diagram of electric domains with spontaneous polarization within a ferroelectric material. The arrows show the directions of the electric polarization vectors.

15.4 FERROELECTRIC MATERIALS

Can materials be spontaneously polarized?

Ferroelectricity is a property of certain dielectrics which exhibit electrical polarization in the absence of an applied electric field. On the scale of a few atoms, ferroelectrics consist of structural units with tiny localized electric dipoles. These dipoles are grouped locally within domains which have spontaneous polarization, as shown in Fig. 15.3, in the same way that a ferromagnet is spontaneously magnetized within domains. The bulk polarization can nevertheless be zero, since the polarizations of all the domains can sum vectorially to zero.

Ferroelectric materials have a macroscopic electric polarization P which is both large and non-linear as a function of electric field ξ. In particular the variation of P with ξ exhibits hysteresis as shown in Fig. 15.4. The name ferroelectric was derived from the name ferromagnet because both show hysteresis, not because of any association between ferroelectrics and the element iron. Ferroelectricity was first discovered in Rochelle salt by Valasek [14,15].

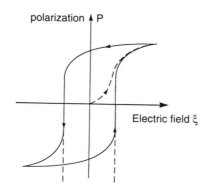

Fig. 15.4 Hysteresis in the dependence of electric polarization on electric field in ferroelectrics.

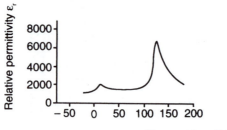

Fig. 15.5 Variation of relative permittivity of barium titanate with temperature. Reproduced with permission of N. Braithwaite and G. Weaver, *Electronic Materials*, published by Butterworths, 1990.

The early theory of ferroelectricity was developed by Kurchatov [16] and was analogous to the classical Weiss mean field theory of ferromagnetism.

Examples of ferroelectric materials include barium titanate, lead zirconate titanate and Rochelle salt (potassium sodium tartrate tetrahydrate).

Typically ferroelectrics have a high relative dielectric constant (relative permittivity) ε_r. A good example is barium titanate $BaTiO_3$. The variation of its relative permittivity with temperature is shown in Fig. 15.5. This can reach a value of several thousand near the Curie point.

15.4.1 Electric polarization in dielectric materials

How can an electric polarization be produced in a material?

An electric polarization can be produced in certain types of materials in two ways. The polarization can arise spontaneously within domains, so that the material forms a low symmetry structure at temperatures below a transition temperature known as the Curie point. Alternatively the polarization can be caused by the application of stress, in which the electric charges in a non-centrosymmetric crystal can be displaced by the stress. In the first case we have a ferroelectric material: in the second case a piezoelectric material.

15.4.2 Depolarized ferroelectrics

Under what conditions does a ferroelectric have the largest differential permittivity in zero electric field?

When a ferroelectric with high relative permittivity ε_r is used in a capacitor it is used in the depolarized condition. From this state the differential permittivity $(= \mathrm{d}P/\mathrm{d}\xi)$ is larger than in the remanent polarized condition and therefore for a given amplitude of electric field ξ the change in polarization P is greater, since $P = \varepsilon_0 \varepsilon_r \xi$.

15.4.3 Domains and domain walls

How is the polarization process explained in terms of changes in the domain configuration?

In materials such as barium titanate, the difference between directions of polarization on either side of the domain walls can be 180° or 90° because the crystallographic directions of polarization are restricted, by crystalline anisotropy, to be along the [100] family of directions as shown in Fig. 15.6.

Since these directions of polarization have associated with them a spontaneous strain, the strain energy is a very significant factor in determining the energy of the domain wall. Consequently it is relatively easy to move a 180° domain

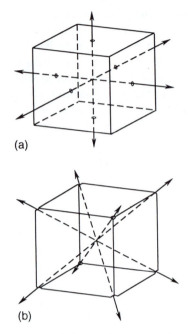

(a)

(b)

Fig. 15.6 Tetragonal and rhombohedral crystal structures, both of which occur in barium ferroelectrics. In the tetragonal phase (a) the ferroelectric domain walls must be either 90° or 180°. In the rhombohedral structure (b) the domain walls can be either 180°, 71° or 109°.

wall under the action of a field because it does not result in any change in strain. However 90° walls are difficult to move because successive regions of the material need to change shape to accommodate the movement. This requires higher energy to overcome obstructions to the necessary deformation from neighbouring grains which must be strained to accommodate the local change in shape. These strains can even be large enough in some cases to cause plastic deformation or even cause the material to fracture.

The high relative differential permittivity of depolarized ferroelectrics arises from the motion of 180° domain walls which are easy to move. At higher polarizations it becomes necessary to move more 90° walls which require higher energy because of the associated change in strain. This leads to a lower differential permittivity. At the Curie temperature and above, the ferroelectric domains are destroyed and the material becomes 'paraelectric'.

15.4.4 Paraelectric phase

What happens to ferroelectrics at higher temperatures?

This phase occurs at higher temperatures, above the Curie point. The Curie point itself depends on chemical composition of the material. In the paraelectric phase polarization is induced by application of an electric field, but when the field is removed it reverts to the unpolarized condition. In this phase the electric dipole vectors are disordered in the absence of an applied field. This is analogous to the behaviour of the magnetization of a paramagnet when subjected to a magnetic field.

15.4.5 Ferroelectric phase

What happens when the disordered paraelectric is cooled below its Curie temperature?

Above the Curie temperature, since the electric dipoles are randomly oriented, no electric domains can exist. Once the material has been cooled below its Curie temperature it becomes ferroelectric with localized spontaneous polarization within domains as shown in Fig. 15.3. The ferroelectric phase is characterized by a very high relative permittivity ε_r. All ferroelectric materials exhibit a paraelectric phase at higher temperatures.

15.4.6 Antiferroelectric phase

Is there an ordered polarized phase that is analogous to antiferromagnetism' with antiparallel alignment of polarizations?

Compositions within 10% of pure $PbZrO_3$ are orthorhombic and below the Curie temperature are antiferroelectric, which means that the neighbouring unit cells of the crystal have polarizations in opposite directions. This is analogous to the antiferromagnetic phase in magnetic materials.

15.5 FERROELECTRICS AS TRANSDUCERS

Which materials are most widely used as transducers?

Ferroelectric materials are particularly useful in transducers. All insulators, whether ferroelectric or not, exhibit electrostriction to a greater or lesser degree. This results directly from the polarization of the material. Since ferroelectrics give much higher polarizations for a given field strength we might expect that these materials will, on the whole, exhibit higher electrostrictive strains. This of course is not a totally general result, but the materials with the highest electrostrictions are also ferroelectrics. These materials, such as barium titanate, are therefore widely used as transducers.

Other ferroelectric materials such as lead zirconate titanate and related materials (PZT), lithium niobate, lead germanate, potassium dihydrogen phosphate, strontium barium niobate have ranges of Curie temperature up to 1500 K and spontaneous polarizations of up to 4 coulombs per square metre.

15.5.1   Piezoelectricity in ferroelectrics

Does a ferroelectric material exhibit electrostrictive or piezoelectric strain under the action of an electric field?

In the 'poled' state, the action of a stress on a ferroelectric produces a change in electric voltage. These materials can therefore act as generators of electric fields. The 'poled' state of the material is one in which it has been subjected to an electric field and left with a remanent polarization. This is equivalent to remanent magnetization in a ferromagnet. The applied stress causes a rearrangement of the electric domains. This changes the bulk polarization, leading to a change in voltage across the material.

Note that while all ferroelectrics can exhibit piezoelectricity (according to the conventional definition as described in section 15.3.6), not all piezoelectric materials are ferroelectrics. For example quartz is piezoelectric but not ferroelectric.

15.5.2   Polycrystalline transducer materials

How does the energy conversion efficiency vary from single-crystal to polycrystalline materials?

The above discussion of electrostriction/piezoelectricity needs to be modified in the case of polycrystalline materials because the change in shape of the grains is opposed by their neighbouring grains. These neighbouring grains will not in general need to reorient, or change shape, at the same field strength. The mechanical work done internally as a consequence of these strains results in lower energy conversion efficiencies.

The use of ferroelectric materials has allowed polycrystalline materials to be employed as transducers. In the early days of piezoelectricity the connection between piezoelectricity and crystal symmetry was so strongly established that the analogy between electrostrictive strain of a ferroelectric and the magnetostrictive strain of a ferromagnet was difficult to recognize. Therefore the piezoelectrics which received attention were crystalline materials without centres of symmetry. Once it was established that the polarity needed to impart piezoelectric properties could be achieved by the temporary application of a strong electric field to ferroelectric materials, a whole range of new polycrystalline piezoelectric/electrostrictive materials became available [17].

15.5.3   Ageing

Does the bulk polarization of a ferroelectric change with time?

A very slow change in remanent polarization occurs with time in ferroelectrics. This results in a decay of remanence through relaxation of the material to its global energy minimum which in zero field occurs at zero bulk polarization P. As in ferromagnets the magnetic domain walls are pinned by defects and this increases both the coercivity and the ageing rate of the material by making changes in polarization arising from domain wall motion more difficult.

15.5.4 Lead zirconate titanate (PZT)

What is the most widely used ferroelectric material?

An important category of ferroelectrics is based on lead, zirconium and titanium oxides. These may be considered the archetypal ferroelectric materials. Suitable adjustment of the chemical composition and microstructure of these materials can result in a wide range of possible properties suitable for many technological needs. In this respect these ceramics play a role in ferroelectrics that is similar to the role of iron in ferromagnets.

The perovskite crystal structure which occurs in barium titanate also occurs in the alloys of the system $PbZrO_3$–$PbTiO_3$, lead zirconate titanate. By varying the ratio of Zr to Ti the properties can be engineered to meet many transducer specifications. The phase diagram of PZT is shown in Fig. 15.7.

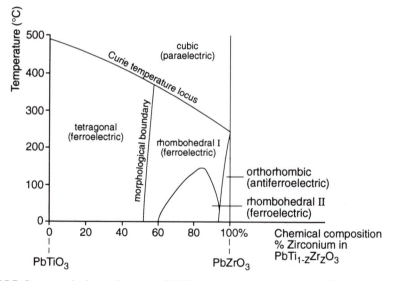

Fig. 15.7 Structural phase diagram of PZT – lead zirconate titanate. Reproduced with permission of N. Braithwaite and G. Weaver, *Electronic Materials*, published by Butterworths, 1990.

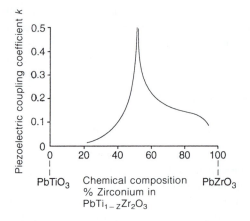

Fig. 15.8 Variation of the coupling coefficient k of PZT with chemical composition. Reproduced with permission of N. Braithwaite and G. Weaver, *Electronic Materials*, published by Butterworths, 1990.

15.5.5   Chemical additions to PZT

How can the piezoelectric properties of PZT be controlled by selection of chemical composition?

The properties of PZT can be strongly influenced by the addition of other metal oxides. These are used to manipulate properties such as conductivity, electric coercivity and elastic modulus. In the PZT compounds the energy conversion efficiency also depends on the chemical composition. If the composition is close to the rhombohedral–tetragonal transition which occurs at about 50% Zr, 50% Ti then the energy conversion efficiency rises to about 25% ($k = 0.5$), from values in the range of 2% ($k = 0.15$) on either side of this composition, as shown in Fig. 15.8. Values of 50% efficiency ($k = 0.7$) have been achieved in some cases.

The reason for this high conversion efficiency at 50% Zr–50% Ti composition is that the grains of the material can find shear transformations between both rhombohedral and tetragonal crystal classes because the energy is finely balanced between these two crystal structures at this composition. These shear transformations therefore occur, as well as the conventional electric polarization rotations, within either the rhombohedral or tetragonal crystal grains. This allows the rotations to proceed more easily via two mechanisms and results in smaller internal energy losses.

The PZT class of ferroelectrics is very important because of the diversity of properties displayed by these compounds. The material properties can be engineered by chemical additions and control of microstructure. This diversity of ferroelectric properties makes the materials very versatile and of crucial importance to the subject of ferroelectricity.

15.5.6 Ferroelectrics for data storage

What are the prospects for using ferroelectric media instead of magnetic media for recording applications?

In much the same way that ferro- and ferrimagnetic materials have been used for data storage, ferroelectrics could also be used in principle, and in fact they have come under consideration for this from time to time. In the past considerable effort has been devoted to this, as indicated by Elliott and Gibson [18, p. 88], but the power input needed to obtain fast switching speeds has been a limitation.

More recently there has been renewed interest by Chikarmane *et al.* [19] and Scott *et al.* [20]. It seems from this work that thin films of lead zirconate titanate may have applications in dynamic random access memories because of the large charge storage density per unit area ($19.6\,\mu C\,cm^{-2}$) and low leakage current density ($1.32 \times 10^{-7}\,A\,cm^{-2}$). The remanent polarization of lead zirconate titanate is typically $10\,\mu C\,cm^{-2}$.

For non-volatile random access memories Ramesh *et al.* [21] have studied heterostructures composed of layers of lead zirconate titanate (PZT) and yttrium barium copper oxide (YBCO). The principal advantage of ferroelectrics over ferromagnetic materials for data storage lies in the high switching speeds which can be achieved. Ferroelectrics can respond to applied fields at frequencies up to the megahertz range, while magnetic materials, particularly electrically conducting magnetic materials, can only respond at kilohertz frequencies. On the other hand, the access time for data storage and retrieval is not limited by the response time of the storage medium alone, and this means that other factors limiting the response time need to be improved before the speed of the magnetic material itself becomes the critical parameter.

REFERENCES

1. Curie, J. and Curie, P. (1880) *Compt. Rend. Acad. Sci.*, **91**, 294.
2. Lines, M. E. and Glass, A. M. (1977) *Principles and Applications of Ferroelectrics and Related Materials*, Clarendon Press, Oxford.
3. Lippmann, G. (1881) *Compt. Rend. Acad. Sci.*, **92**, 1049.
4. Kumar, U., Halliyal, A. and Cross, L. E. (1986) *IEEE Sixth Symposium on Applications of Ferroelectrics, Lehigh University Pennsylvania*, IEEE Press, p. 633.
5. Mitsui, T., Tatsuzaki I. and Nakamura, E. (1976) *An Introduction to the Physics of Ferroelectrics*, Gordon and Breach Publishers, London, p. 77.
6. Uchino, K. (1986) *IEEE Sixth Symposium on Applications of Ferroelectrics, Lehigh University, Pennsylvania*, IEEE Press.
7. Newnham, R. E., Xu, Q. C., Kumar, S. and Cross, L. E. (1990) *Ferroelectrics*, **102**, 259.
8. Joule, J. P. (1842) *Ann. Electr. Magn. Chem.*, **8**, 219.
9. Cullity, B. D. (1971) *J. Met.* **23**, 35.
10. Clark A. E. (1980) Magnetostrictive rare earth iron compounds, in *Ferromagnetic Materials*, vol. 1, (ed. E. P. Wohlforth), North Holland Publishing, Amsterdam.

11. Jiles, D. C. (1990) Development and characterization of the highly magnetostrictive alloy Tb-Dy-Fe for use in sensors and actuators, in *New Materials and their Applications*, Institute of Physics Publishing, Bristol.
12. Schieber, M. M. (1967) *Experimental Magnetochemistry*, John Wiley & Sons, New York, pp. 55 and 401.
13. Jiles, D. C. and Atherton, D. L. (1984) *J. Phys. D. (Appl. Phys.)*, **17**, 1265.
14. Valasek, V. (1920) *Phys. Rev.*, **15**, 537.
15. Valasek, V. (1921) *Phys. Rev.*, **17**, 475.
16. Kurchatov, I. V. (1933) *Segnetoelektriki*, CM-GTTI, Moscow.
17. Jaffe, B., Cook, W. R. and Jaffe, H. (1971) *Piezoelectric Ceramics*, Academic Press.
18. Elliott, R. J. and Gibson, A. F. (1974) *An Introduction to Solid State Physics and Its Applications*, MacMillan, London.
19. Chikarmane, V. *et al.* (1972) *J. Electr. Mater.*, **21**, 503.
20. Scott, J. F., Araujo, C. A., Brett Meadows, H. *et al.* (1989) *J. Appl. Phys.*, **66**, 1444.
21. Ramesh, R. *et al.* (1992) *J. Electr. Mater.*, **21**, 513.

FURTHER READING

Blinc, R. and Zeks, B. (1974) *Soft Modes in Ferroelectrics and Antiferroelectrics*, North Holland.
Braithwaite, N. and Weaver, G. (1990) *Electronic Materials*, Butterworths, London, Chapter 4.
Burfoot, J. C. and Taylor, G. W. (1979) *Polar Dielectrics and their Applications*, MacMillan, London.
Cady, W. G. (1964) *Piezoelectricity*, Dover Publications, New York.
Fridkin, V. M. (1980) *Ferroelectric Semiconductors*, Consultants Bureau, New York.
IEEE Standard on Piezoelectricity (1988) Standard No. 176–1987, IEEE, New York.
Jaffe, B., Cook, W. R. and Jaffe, H. (1971) *Piezoelectric Ceramics*, Academic Press.
Lines, M. E. and Glass, A. M. (1977) *Principles and Applications of Ferroelectrics and Related Materials*, Clarendon Press, Oxford.
Mason, W. P. (1950) *Piezoelectric Crystals and their Applications to Ultrasonics*, Van Nostrand, New York.
Newnham, R. E. (1989) Electroceramics, *Rep. Prog. Phys.*, **52**, 123.
Rosen, C. Z., Hiremath, B. V. and Newnham, R. E. (1992) *Piezoelectricity*, Key papers in Physics Series, American Physical Society, New York, 1992.
Proceedings of the Sixth IEEE International Symposium on Applications of Ferroelectrics, Lehigh University, Bethlehem, Pennsylvania, IEEE Press, 1986.

16

Electronic Materials for Radiation Detection

In this final chapter we look at the use of electronic materials for the detection of ionizing radiation. This is another specialized application in which the electronic structures of semiconductor materials are used to convert incident radiation into electrical voltages which can be easily measured and characterized. In fact the radiation sensors are really just a different form of transducer which converts radiative energy into electrical energy. Four main types of sensor are discussed: ionization chambers, semiconductors, scintillation detectors and thermoluminescent dosimeters. The principles underlying the operation of each of these are discussed and the particular applications are described.

16.1 RADIATION SENSORS

What types of sensor are available for the detection of radiation?

Radiation detectors are forms of transducers since they convert the energy of incident radiation into electrical voltages which can be measured and analysed through electronic instrumentation. They therefore form the interface between the measuring equipment and the radiation which it is designed to measure. The main class of detectors is actually a form of photoelectric transducer which we have already met as the photodetector in our discussion of semiconductor devices in section 12.1. A second group is formed by the scintillation detectors. Other forms of radiation sensor are the ionization detectors such as ionization chambers, Geiger–Muller tubes and other gas- or liquid-filled detectors, and thermoluminescent dosimeters.

We are interested here in the detection of radiation, that is, α, β, γ and neutrons. All methods of detecting nuclear particles depend on the generation of a voltage as the radiation passes through the detector or sensor. If the radiation consists of a charged particle as in α and β radiation, then this ionization is immediately obvious. However, if the radiation is a neutral particle, for example a neutron, it will not cause ionization directly and therefore an intermediate process is necessary in which an energetic charged particle is produced.

There is a variety of different radiation detectors. We shall classify these detectors as follows:

- gas-filled detectors
- semiconductor detectors
- scintillation counters
- thermoluminescent dosimeters.

16.2   GAS-FILLED DETECTORS

How can ionization of a gas be used to detect radiation?

The earliest type of radiation detector was one in which the ionizing material was a gas. This type of detector provides a basic understanding of the interaction of radiation with materials which is common to all forms of radiation sensor. It therefore deserves discussion first. Nevertheless, from the standpoint of this book the semiconductor and scintillation detectors, which we consider later, are closer to our main theme which is applications of electronic materials.

The best known of the gas-filled detectors is the Geiger–Muller tube (Fig. 16.1). However, strictly speaking, this really only refers to a mode of operation which applies equally well to liquid- or gas-filled detectors. The principle of operation of the sensor is that ionization takes place in an enclosed chamber in the presence of an electric field. The voltage gradient of the field causes the negative ions to be swept towards the anode and the positive ions to be swept towards the cathode. This is identical to the effect of a potential gradient in a p-n junction. If the two electrodes are then connected in some way through a load resistor, the ionization event leads to a current pulse which can be used as a means for detecting the number of ionizing particles passing through the chamber, and measuring their energy.

In its most common form, the detector consists of a cylindrical tube with an outer cylindrical cathode and a central anode. These detectors can be operated in three different modes: a) ionization chamber, b) proportional counter and c) Geiger–Muller mode. The important point is that the gas in the chamber should be easily ionized by incident radiation. Usually gases such as

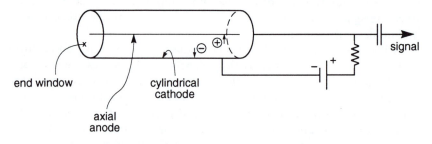

Fig. 16.1 Schematic diagram of the main features of a gas-filled detector.

helium, argon, or neon with a small amount of ethyl alcohol or ethyl bromide are used. A clear and concise description of this type of detector has been given by Kleinknecht [1].

16.3 SEMICONDUCTOR DETECTORS

How can the semiconductor photodetectors be used for detection of incident radiation?

Semiconductor detectors, also known as solid-state radiation detectors, function as solid ionization chambers. These have several advantages over the gas-filled detectors because they can be made compact and rugged. For example, the density of a solid-state detector is typically 1000 times the density of gas in a gas-filled detector. Therefore to a first approximation the semiconductor detector could be one thousandth of the volume of the equivalent gas-filled detector.

The semiconductor detectors became widely available in the 1960s. They are based on the principle of electron–hole pair generation and collection. In addition to their compact size they are very fast in their operation and have a wide range of applications including detection of both photons and charged particles. They also have some drawbacks which include a predisposition to degradation from radiation-induced damage.

We have already discussed one form of semiconductor radiation detector in detail – the photodetector. In this device, the electron–hole pair was created by an incident light photon whose energy was greater than or equal to the band gap (Fig. 16.2).

16.3.1 Semiconductor junction detectors

How does the semiconductor junction detector work?

A junction between pieces of n-type semiconductor and p-type semiconductor can also be used as a radiation detector. Usually the junction is reverse biased, as in the photodetector. The existence of the depletion layer (space charge region) is the key to the use of the p-n junction as a radiation detector. When ionization occurs in the depletion layer, the electrons are elevated to the conduction band. As a result of the potential difference across the depletion layer they are swept out of this region giving a current pulse for each ionization event.

16.3.2 Fabrication of semiconductor radiation detectors

How exactly are semiconductor detectors formed?

Two types of semiconductor junction diode detector that have been in widespread use are the surface barrier detector and the diffused-junction detector. These

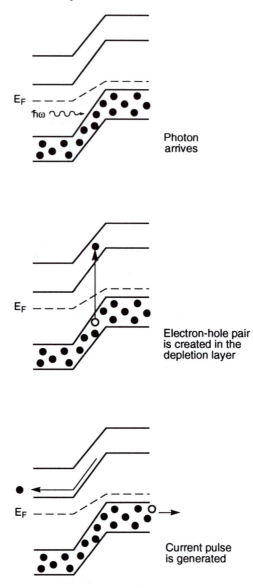

Fig. 16.2 Excitation of an electron–hole pair by incident radiation giving a current pulse which can be used to detect the radiation. Note, that in order to detect the current pulse, a field gradient must be present.

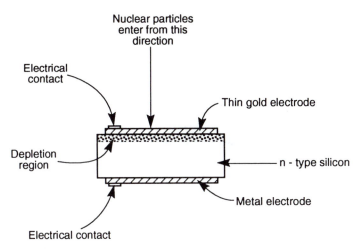

Fig. 16.3 Surface barrier detector. Reproduced with permission of W. Price, *Nuclear Radiation Detection*, published by McGraw-Hill, 1964.

are semiconductor junction detectors which have undergone surface modification to enhance their efficiency for radiation detection.

The surface barrier detector is a slab of n-type semiconductor on which a p-type layer has been formed by oxidation. A thin metallic coating is applied to each face to act as an electrode (Fig. 16.3). Typically, gold is used as the electrode on the p-type material (cathode) and aluminium is used as the electrode on the n-type material (anode). This type of detector was developed first by Dearnaley and Whitehead [2]. A recent description of the fabrication of surface barrier detectors using high-purity single-crystal silicon has been given by Shiraishi *et al.* [3].

In order to function properly, the ionization process must occur in the depletion layer where the electron bands have a high potential gradient, giving a strong local electric field which accelerates the electrons. This means that (i) the thickness of the depletion layer should be greater than the range of the incident particles in the material and (ii) the depletion layer must extend to the surface of the material.

The diffused-junction detector consists of a slab of p-type material, for example p-type silicon, on one surface of which donor atoms, such as phosphorus, have been diffused to produce an n-type semiconducting layer. A layer of aluminium is then evaporated on the opposite side of the silicon from the phosphorus (Fig. 16.4). The original construction and operation of the diffused junction detector has been described by Moncaster *et al.* [4].

The depth of the depletion layer which arises naturally from a p-n junction in silicon is typically 30 μm. This is not adequate for most radiation-detection applications because the range of the ionizing radiation in the material is greater than this depth. However, the depth can be extended by the application of a

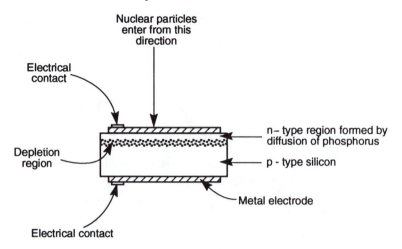

Fig. 16.4 Diffused-junction detector. Reproduced with permission of W. Price, *Nuclear Radiation Detection*, published by McGraw-Hill, 1964.

reverse external bias, hence the name diffused-junction is used to describe this detector. Recent developments in diffused-junction detectors have been described by Walton and Haller [5].

16.3.3 Lithium-drifted germanium (GeLi) detectors

How can the active volume of a semiconductor detector be increased?

We know that in order to function properly the ionization events need to occur in the depletion layer of the detector where there is a potential gradient. But there is a limit to the depth of the depletion layer which can be obtained by reverse biasing of a surface barrier or diffused-junction detector. This arises because, as we already know, if the external voltage is too high the dielectric strength of the material is exceeded, the semiconductor junction breaks down and the reverse current increases rapidly. This breakdown voltage has been described in Chapter 11 (see for example Fig. 11.6). Consequently, such detectors have rather limited uses if the radiation has a long range in the semiconductor, as in the case of beta and gamma radiation.

One solution is to produce a 'compensated' region in which the numbers of donors and acceptors are equal. If this is sandwiched between regions of p-type and n-type semiconductor, the holes and electrons migrate away from this region leaving it carrier-free and neutral. If electron–hole pairs are produced in this region, they are swept away giving a current pulse which can be used to count the incoming radiation. This type of detector was first described by Mayer [6].

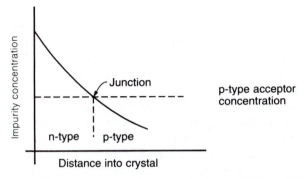

Fig. 16.5 Variation of lithium concentration with depth in a GeLi or SiLi detector after diffusion.

16.3.4 Fabrication process for Li-drifted GeLi detectors

How is the compensated region extended?

The compensated region can be extended by diffusing an electron donor such as lithium into p-type semiconducting germanium or silicon. Initially, the lithium is diffused into p-type material so that the lithium concentration decays exponentially with depth, as shown in Fig. 16.5.

The material is then heated to about 400 °C to aid the diffusion. In those regions in which the lithium concentration is greater than the acceptor concentration, the material becomes n-type. A reverse bias is then applied at a temperature of about 40 °C so that the lithium will drift towards the negative electrode that is down the field gradient. Eventually the lithium concentration as a function of distance varies as shown in Fig. 16.6.

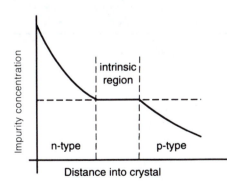

Fig. 16.6 Variation of lithium concentration with depth in a GeLi or SiLi detector after 'drifting'.

It can be seen that there is now a large region where the donor and acceptor levels are equal. This is known as the 'compensated' or intrinsic region. Any local regions in which the lithium-ion concentration is not equal to the acceptor concentration, lead to the existence of a local potential which equalizes the concentrations, and therefore the lithium- and acceptor-ion concentrations are maintained equal throughout this region.

The germanium-lithium detector is used exclusively for gamma ray detection since the germanium has a higher atomic number than silicon, and is therefore more efficient at absorbing the gamma rays. The related silicon-lithium detector is used for beta particle detection. Because of the small band gap in germanium (0.7 eV) the GeLi detector is operated at liquid nitrogen temperatures in order to reduce the dark current. The dark current has been discussed in section 7.5 (see Fig. 7.13 for example). Silicon, with its larger band gap (1.1 eV), has a much smaller dark current at room temperature.

The main advantage of these detectors over the surface barrier and diffused-junction detectors is the length of the transition region between the n-type and p-type semiconductor. Since the radiation needs to deposit all of its energy in this region, for the detector to be useful the simple surface barrier and diffused-junction detectors are only useful for radiation with short path lengths in the material, such as alpha particles and some low-energy beta particles. The GeLi and SiLi detectors with larger junction regions are much better suited for the detection of more penetrating radiations such as gamma and high-energy beta.

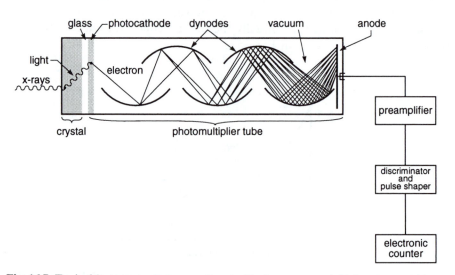

Fig. 16.7 Typical 3-stage arrangement of a scintillation counter. A high voltage difference is applied across each stage of the photo-multiplier which accelerates the electrons and leads to amplification by increasing the number of electrons at each photomultiplier stage.

16.4   SCINTILLATION DETECTORS

How are luminescent materials used for radiation detection?

The principle behind the scintillation detector is that in some materials incident ionizing radiations cause the generation of light which is then detected using a photosensitive sensor. Typical scintillation materials include sodium iodide, lithium iodide, caesium iodide, certain organic phosphors and plastics and even liquid scintillators. These detectors are mainly used for detection of gamma rays and neutrons but can also be used for low-energy beta particles. A typical arrangement for such a detector is shown in Fig. 16.7.

The incident radiation causes the generation of a photon in the scintillator. The photocathode detects the incident light photon and converts it into an

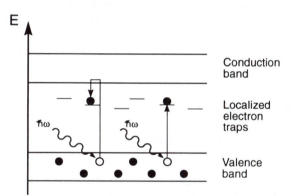

a) Incident photon causes generation of exciton (electron-hole pair) either directly or via the conduction band

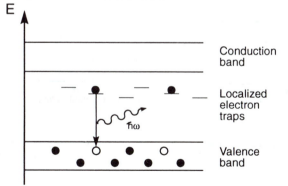

b) Subsequent recombination of the electron-hole pair causes emission of light.

Fig. 16.8 Band structure diagram of a scintillation detector. Incident radiation causes ionization into an electron–hole pair, and subsequent recombination causes emission of light.

electrical signal in the form of a free electron. The photomultiplier stage is a high-gain, low-noise device which amplifies the signal. The traditional photomultipliers used in scintillators have now been replaced by high-gain solid-state photodiodes [7].

16.4.1   Scintillator: principles of operation

How is light produced in a scintillator by high-energy incident radiation?

The principles of operation of scintillators have been described in detail by Birks [8]. The scintillator is simply a luminescent material (section 9.4). The initial excitation which leads to the generation of an electron–hole pair can come from many origins but here we are concerned only when this occurs as a result of incident ionizing radiation. When the electron–hole pair recombines a light photon is emitted. This is detected by the photodetector (photocathode) which is a semiconductor light detector of the type which we have discussed before in section 12.1.1. This converts light into electrons for amplification by the associated instrumentation.

16.5   THERMOLUMINESCENT DETECTORS

How can thermoluminescent materials be used for measurements of total absorbed radiation?

We have discussed the principles of operation of thermoluminescent materials in detail in section 8.6. The electron traps in the band gap of semiconducting materials can be filled by excitation of electrons from the valence band on

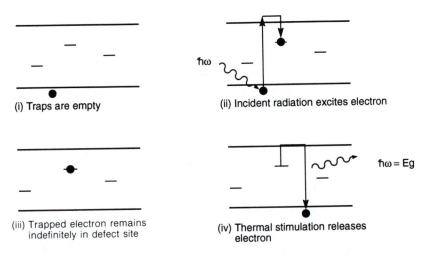

(i) Traps are empty

(ii) Incident radiation excites electron

(iii) Trapped electron remains indefinitely in defect site

(iv) Thermal stimulation releases electron

Fig. 16.9 The four stages of thermoluminescence: (i) initial state; (ii) radiation absorption; (iii) trapped state and (iv) thermal release of electron with light emission.

absorption of incident radiation. The traps can then be emptied when the material is heated and the subsequent decay of electrons into the valence band emits light whose wavelength is characteristic of the band gap energy.

The emitted intensity of light as a function of temperature (the glow curve) can be used to measure the total dose of radiation absorbed by the thermoluminescent material. The intensity of the glow curve at different locations can be calibrated with the amount of absorbed radiation. Once the material has been heated the electron traps are emptied and it is effectively 'reset' to zero. That is, the 'memory' of the previous dose of radiation is erased.

Subsequent measurement of the thermoluminescence glow curve is indicative of the dose since the material was last heated and 'reset'. This technique can also be used for archaeological dating of materials as indicated by Zimmermann [9]. A review of thermoluminescent detectors has been given by Bull [10]. Other methods for detection of radiation include solid-state track detectors as described by Durrani and Bull [11].

16.6 PYROELECTRIC SENSORS

How do heat sensors work?

This class of transducer converts heat radiation into a voltage. Since heat radiation consists of long-wavelength electromagnetic waves, which are located in the infra-red region of the spectrum, these transducers can be thought of as a special case of the photoelectric transducers. However their mechanism is somewhat different. The pyroelectric transducers rely on a change in electric polarization which results from changes in temperature when the material is heated by infra-red radiation. This causes a change in voltage in the sensor by altering the charge per unit area on the surface of the dielectric and this voltage can be detected by ancillary electronic instrumentation such as a voltmeter.

REFERENCES

1. Kleinknecht, K. (1986) *Detectors for Particle Radiation*, Cambridge University Press.
2. Dearnaley, G. and Whitehead, A. B. (1961) *Nucl. Instr. & Methods*, **12**, 205.
3. Shiraishi, F. *et al.* (1983) Thick surface barrier detector made of ultra high purity p type Si single crystal, in *Nuclear Radiation Detector Materials*, (eds E. E. Haller, H. W. Kramer and W. A. Higinbotham), North Holland, New York, p. 175.
4. Moncaster, M. E., Northrop, D. C. and Raines, J. A. (1963) *Nucl. Instr. & Methods*, **22**, 157.
5. Walton, J. T. and Haller, E. E. (1983) Silicon radiation detectors – materials and applications, in *Nuclear Radiation Detector Materials*, (eds E. E. Haller, H. W. Kramer and W. A. Higinbotham), North Holland, New York, p. 141.
6. Mayer, J. W. (1962) *IRE Trans. Nucl. Sci.*, **NS-9**, (3), 124.
7. Reiff, G., Squillante, M. R., Serreze, H. B. and Futine, G. (1983) Large area silicon avalanche photodiodes: a photomultiplier tube alternate, in *Nuclear Radiation Detector Materials* (eds E. E. Haller, H. W. Kramer and W. A. Higinbotham), North Holland, New York, p. 131.

8. Birks, J. B. (1964) *Theory and Practice of Scintillation Counting*, Pergamon Press, Oxford.
9. Zimmermann, D. W. (1971) *Science*, **174**, 818.
10. Bull, R. K. (1986) *Nuclear Tracks Radiat. Meas.* **11**, 105.
11. Durrani, S. A. and Bull, R. K. (1987) *Solid State Nuclear Track Detection*, Pergamon, Oxford.

FURTHER READING

Price, W. J. (1964) *Nuclear Radiation Detection*, McGraw Hill, New York.
Snell, A. H. (1964) *Nuclear Instruments and their Uses*, John Wiley & Sons, New York.

Solutions

Exercise 1.1 *The Wiedemann–Franz law.* This states that the thermal and electrical conductivities of metals are related by the equation,

$$\frac{K}{\sigma T} = L$$

where K is the thermal conductivity in $J\,m^{-1}\,K^{-1}\,s^{-1}$, σ is the electrical conductivity in $\Omega^{-1}\,m^{-1}$ and T is the temperature in degrees Kelvin. L is the Lorentz number which has the value $2.4 \times 10^{-8}\,J\,\Omega\,K^{-2}\,s^{-1}$

Exercise 1.2 *The Hagen–Rubens relation.* This relates the optical reflectivity and electrical conductivity of a material. The relation is only valid for longer wavelengths, that is in the infra-red region of the spectrum ($v < 10^{14}\,Hz$, $\lambda > 3\,\mu m$),

$$R = 2\sqrt{\frac{4\pi\varepsilon_0 v}{\sigma}}$$

where R is the reflectance, which is dimensionless, v is the frequency in Hz and σ is the electrical conductivity in $\Omega^{-1}\,m^{-1}$.

Exercise 1.3 *The Dulong–Petit law.* This law relates the specific heat capacity of a solid to the number of atoms. It was originally noted that the molar heat capacity of solids was similar for a large number of materials, being typically $25\,J\,mol^{-1}\,K^{-1}$. Later this was explained in terms of the vibrational modes of the atoms within the solid. If each atom has three degrees of freedom and behaves as a harmonic oscillator with energy $k_B T$ along each of these degrees of freedom, then the internal energy of the solid will be

$$U = 3Nk_B T$$

where N is the number of atoms per unit volume, k_B is Boltzmann's constant and T is the absolute temperature. If we consider a mole of the material, then $N = N_0 = 6.025 \times 10^{23} =$ Avogadro's number and the molar heat capacity $C = dU/dT$, is

$$C = 3N_0 k_B$$
$$= 24.96 \, \text{J mol}^{-1} \, \text{K}^{-1}.$$

Exercise 2.1   Elastic modulus of linear atomic lattice. In this situation, there is an equation for the potential energy with two unknowns and two pieces of information which allow the unknowns to be calculated

$$E_P = \alpha_1 a^{-9} - \alpha_2 a^{-1}.$$

At equilibrium separation a_0 we must have the derivative of energy with respect to displacement equal to zero, $(dE_P/da)_{a_0} = 0$. Therefore,

$$0 = -9\alpha_1 a_0^{-10} + \alpha_2 a_0^{-2}$$

with $a_0 = 0.3 \times 10^{-9}$ m. The dissociation energy is the difference in energy between this equilibrium position, and complete separation of atoms at $a = \infty$. Therefore setting $E(\infty) = 0$ and,

$$E_P(a_0) = -4 \, \text{eV}$$
$$= -6.4 \times 10^{-19} \, \text{joules}$$

we obtain,

$$-6.4 \times 10^{-19} = \alpha_1 a_0^{-9} - \alpha_2 a_0^{-1}$$

with $a_0 = 0.3 \times 10^{-9}$ m. This gives two simultaneous equations in the two unknowns. Solving these for α_1 and α_2 gives

$$\alpha_1 = 1.57 \times 10^{-105} \, \text{J m}^9$$
$$\alpha_2 = 2.16 \times 10^{-28} \, \text{J m}.$$

If we consider the elastic modulus E_Y as the derivative of the applied force F_{app} per unit area with respect to the strain e then in the case of a linear lattice the concept of 'pressure' is not very meaningful. Suppose we consider the effective cross-sectional area per linear lattice chain to be a_0^2 (i.e. equivalent to a simple cubic lattice), then the stress becomes equal to F/a_0^2, and therefore the elastic modulus is

$$E_Y = \frac{1}{a_0^2} \left(\frac{dF_{app}}{de} \right)$$

and since the strain e can be represented as $(a - a_0)/a_0$ where a_0 is the equilibrium separation of the atoms, and a is the separation at strain e

$$\frac{de}{da} = \frac{1}{a_0}.$$

Therefore,

$$E_Y = \frac{1}{a_0} \left(\frac{dF_{app}}{da} \right)_{a_0}$$

and $F_{app} = -F_{internal} = dE_P/da$, so

$$E_Y = \frac{1}{a_0}\left(\frac{d^2 E_P}{da^2}\right)_{a_0}$$

$$= \frac{1}{(0.3 \times 10^{-9})}(90\alpha_1 a_0^{-11} - 2\alpha_2 a_0^{-3})$$

$$E_Y = 0.213 \times 10^{12}\ \text{Pa}.$$

The force on the lattice is related to the energy $E_P(a)$ by the equation,

$$F = -\left(\frac{dE_P(a)}{da}\right)_{a_1}$$

where $a_1 = 0.99\ a_0 = 0.297$ nm is the lattice spacing at 1% strain,

$$F = 9\alpha_1 a_1^{-10} + -\alpha_2 a_1^{-2}$$
$$= 1.92 \times 10^{-10}\ \text{N}.$$

Exercise 2.2 Lattice stabilized by electrostatic repulsion. In this case consider the force on a given atom in the linear lattice due to its interactions with its two nearest neighbours. If x_0 is the equilibrium separation and Δx is the displacement of one atom from equilibrium and q is the charge per ion, then the force on that atom will be,

$$F(\Delta x) = \frac{q^2}{4\pi\varepsilon_0}\left(\frac{1}{(x_0 + \Delta x)^2} - \frac{1}{(x_0 - \Delta x)^2}\right)$$

which from the previous considerations is known to be equal to the negative derivative of the energy E_P with respect to position x, that is $-dE_P/dx$. The elastic modulus $E_Y = x_0(dF/dx)_{x_0}/A$ where A is the cross-sectional area. If we assume a simple cubic lattice then the cross-sectional area per linear lattice chain will be x_0^2. Therefore the elastic modulus is $E_Y = (dF/dx)_{x_0}/x_0$. We have defined the force constant k as,

$$k = \left(\frac{dF}{dx}\right)_{x_0}$$

so that in this case $E_Y = k/x_0$, then from the wave equation in section 2.2 the velocity of longitudinal waves v is given by,

$$v = \sqrt{\frac{kx_0^2}{m}} = \sqrt{\frac{E_Y x_0^3}{m}}$$

differentiating the expression for F gives,

$$k = \left(\frac{dF}{dx}\right)_{x_0} = 4\frac{q^2}{4\pi\varepsilon_0}\frac{1}{x_0^3}$$

giving

$$k = 7.36 \, \text{N m}^{-1}.$$

Now substituting the values of k, x_0 and m into the equation for velocity gives

$$v = 4.15 \times 10^3 \, \text{m s}^{-1}$$

and the elastic modulus is

$$E_Y = 1.47 \times 10^{10} \, \text{Pa}.$$

Exercise 2.3 Classical and Debye theories of specific heat. The Dulong–Petit law states that the specific heat capacity of one mole of material is

$$C = 3N_0 k_B$$

where N_0 is Avogadro's number and k_B is Boltzmann's constant. Since the heat capacity is simply the derivative of the internal energy U with respect to temperature,

$$C = \frac{dU}{dT}$$

therefore the internal energy U is given by

$$U = \int C \, dT$$

$$= 3N_0 k_B T$$

with $T = 300 \, \text{K}$

$$U = 7477 \, \text{joules}.$$

In order to demonstrate that the Debye theory gives the classically expected Dulong–Petit result at high temperature, we can start from the Debye expression for the internal energy,

$$U = \frac{9N k_B T^4}{\theta_D^3} \int_0^{x_{max}} \frac{x^3}{e^x - 1} \, dx$$

where $x = \hbar\omega / k_B T$ and $x_{max} = \theta_D / T$. At high temperatures x becomes small so that

$$e^x - 1 \approx x.$$

Hence,

$$U = \frac{9N k_B T^4}{\theta_D^3} \int_0^{x_{max}} x^2 \, dx$$

$$= \frac{9Nk_B T^4}{\theta_D^3} \left(\frac{x^3}{3}\right)_0^{\theta_D/T}$$

leading to the following result for the internal energy of one mole ($N = N_0$) at high temperature

$$U = 3N_0 k_B T$$

which is the classically expected result.

Using Fig. 2.9 which shows the variation in heat capacity with temperature, the energy is then simply the integral $\int_0^T C_v dT = U(T)$. At $T = 300$ K, $T/\theta_D = 0.70$, where the integral is clearly about half of the classical value of $U(T)$. Therefore

$$U(T) \approx 3750 \text{ joules.}$$

Exercise 3.1 Drude free-electron theory of metals. The Drude theory attempts to explain only those properties of a metal which arise from the electronic properties. These include the relationship between electrical and thermal conductivity known as the Wiedemann–Franz law. In this respect the 'theory' is in fact only a rather limited model of certain restricted properties of metals. The theory describes the electrons as a classical free-particle gas and assumes that they are subjected to a constant background energy and therefore do not 'see' the periodic potential wells associated with the ionic cores.

The main assumptions of the Drude theory are as follows:

1. collisions between electrons are instantaneous and elastic and this is the mechanism by which thermal conductivity takes place;
2. other interactions between the electrons or the electrons and ions can be neglected;
3. the mean free time of electrons between collisions is independent of the electron position and velocity.

Therefore the Drude model gave the equation of motion of the electrons in the metal as,

$$m\frac{d^2 x}{dt^2} + \gamma\frac{dx}{dt} = e\xi$$

where m is the electronic mass, e the electronic charge, ξ the applied electric field and γ is a damping factor, which prevents the electrons from accelerating indefinitely under the action of a field ξ.

The classical Drude free-electron model was extended by Lorentz to include bound electrons because empirical results showed structure in the optical spectra at higher energies which were not interpretable on the basis of the existing Drude theory. Lorentz's idea was that the bound electrons, which were the inner electrons of the atoms forming the metal and hence were not able to move freely throughout the metal, could be described as if they were harmonic oscillators.

The Lorentz theory, by including the bound electrons, arrived at an equation of motion of the form

$$m\frac{d^2x}{dt^2} + \gamma'\frac{dx}{dt} + kx = e\xi$$

for the bound electrons where k is the measure of the binding strength of the electrons to their atoms. The optical properties of a metal could now be described in terms of both types of electron, that is both free and bound. In practice the Lorentz model leads to absorption bands in the reflectivity at higher frequencies.

Exercise 3.2   Reflectivity based on Drude theory. As shown in the chapter, the dielectric constants ε_1 and ε_2 can be derived on the basis of the Drude model in terms of the two frequencies v_1 and v_2

$$\varepsilon_1(v) = n^2 - k^2 = 1 - \left(\frac{v_1^2}{v^2 + v_2^2}\right)$$

$$\varepsilon_2(v) = 2nk = \frac{v_2}{v}\left(\frac{v_1^2}{v^2 + v_2^2}\right)$$

where v_1 is the 'plasma' frequency and v_2 is the 'damping' frequency. From classical optics the reflectance R is related to the dielectric constants by the equations

$$R = \frac{\sqrt{\varepsilon_1^2 + \varepsilon_2^2} + 1 - \sqrt{2(\sqrt{(\varepsilon_1^2 + \varepsilon_2^2)} + \varepsilon_1)}}{\sqrt{\varepsilon_1^2 + \varepsilon_2^2} + 1 + \sqrt{2(\sqrt{(\varepsilon_1^2 + \varepsilon_2^2)} + \varepsilon_1)}}.$$

The predictions of the Drude theory can be found from this equation at very high and very low frequencies. However at intermediate frequencies the evaluation of R is somewhat cumbersome. Nevertheless, starting with the limit as $v \to 0$

$$\lim_{v \to 0} \varepsilon_1 = 1 - v_1^2/v_2^2$$

$$\lim_{v \to 0} \varepsilon_2 = \infty.$$

Consequently, ε_2 dominates in the above equation for reflectance R at low frequencies and therefore,

$$\lim_{v \to 0} R = 1.$$

Secondly, looking at the high frequency limit as $v \to \infty$, it is easily seen that

$$\lim_{v \to \infty} \varepsilon_1 = 1$$

$$\lim_{v \to \infty} \varepsilon_2 = 0$$

consequently the reflectance goes to zero at high frequencies.

$$\lim_{\nu \to \infty} R = 0.$$

A sketch of the value of R as a function of frequency ν for the Drude model is given in Fig. 3.6.

The Drude model for reflectance of metals works reasonably well for low frequencies, specifically in the infra-red range of the spectrum. It fails to account for the optical properties of metals at higher frequencies, in the optical range and beyond.

Exercise 3.3   Electrical and optical properties of a classical free-electron metal. The resistive coefficient γ is related to the mobility of the electrons by the relation

$$\gamma = \frac{e}{\mu}.$$

Therefore inserting the value of μ for copper gives

$$= \frac{1.602 \times 10^{-19}\,\text{C}}{3.5 \times 10^{-3}\,\text{m}^2\,\text{V}^{-1}\,\text{s}^{-1}}$$

$$\gamma = 0.46 \times 10^{-16}\,\text{N}\,\text{s}\,\text{m}^{-1}.$$

The electrical conductivity σ is itself related to the resistive coefficient γ by the equation,

$$\sigma = \frac{n_f e^2}{\gamma} = n_f e \mu$$

and from the density, atomic weight and number of electrons per atom in copper the number of conduction electrons in a cubic metre of copper is $N_f \approx 8.5 \times 10^{28}\,\text{m}^{-3}$. Therefore the electrical conductivity is

$$\sigma = \frac{(8.5 \times 10^{28})(1.602 \times 10^{-19})^2}{0.46 \times 10^{-16}}$$

$$\sigma = 0.476 \times 10^8\,\Omega^{-1}\,\text{m}^{-1}.$$

The time constant between collisions of the electrons τ is

$$\tau = \frac{m}{\gamma}$$

$$= \frac{9.109 \times 10^{-31}\,\text{kg}}{0.46 \times 10^{-16}\,\text{N}\,\text{s}\,\text{m}^{-1}}$$

$$\tau = 19.8 \times 10^{-15}\,\text{s}.$$

Exercise 4.1   Fermi energy for a free-electron metal. The number of states per unit volume of k-space is $1/4\pi^3$ which allows for one spin-up and one spin-down state at each k-state. Assuming a spherically symmetric distribution in k-space because the electrons are by definition completely free, then the region of k-space between k and $k + dk$ has a volume,

$$dV = 4\pi k^2 dk$$

and therefore the number of electron states between k and $k + dk$ is simply this volume multiplied by $1/4\pi^3$

$$N(k)dk = \frac{1}{4\pi^3} 4\pi k^2 dk$$

$$= \frac{k^2}{\pi^2} dk.$$

This can now easily be transformed into an energy expression $N(E)dE$, since for completely free electrons the energy is

$$E = \frac{\hbar^2 k^2}{2m}$$

so $k^2 = 2mE/\hbar^2$, and $dk = (2m/\hbar^2)^{1/2}\frac{1}{2}E^{1/2}dE$.

Therefore, the number of electron states between energies E and $E + dE$ is obtained by substitution into the expression for $N(k)dk$, giving

$$N(E)dE = \frac{1}{2\pi^2}\left(\frac{2m}{\hbar^2}\right)^{3/2} E^{1/2}dE.$$

The total number of electrons per unit volume of space will then be the integral of this expression, and is therefore

$$2N_0(E) = \frac{1}{3\pi^2}\left(\frac{2m}{\hbar^2}\right)^{3/2} E^{3/2} = \frac{k^3}{3\pi^2}.$$

This expression includes all electrons from energy zero up to an energy E. Therefore if the free-electron Fermi energy is E_F it can be related to the total number of electrons per unit volume N_{Tot} $(= 2N_0(E_F))$ by the equation

$$E_F = (3\pi^2 N_{Tot})^{2/3} \frac{\hbar^2}{2m}.$$

It must be remembered that this expression for the Fermi energy only applies to free electrons.

Exercise 4.2   Solution of wave equation in a finite square well. We consider a one-dimensional finite square-well potential of height V_0 and of spatial extent $2a$ as shown in Fig. 4.4, and applying the time-independent Schrödinger wave equation,

$$-\frac{\hbar^2}{2m}\frac{d^2\Psi(x)}{dx^2} + V(x)\Psi(x) = E\Psi(x)$$

which is simply a statement that the kinetic energy plus the potential energy equals the total energy. We must now find the set of wave functions $\Psi_i(x)$ which will satisfy this equation with the given boundary conditions. Note that without boundary conditions all wave functions $\Psi(x)$ will satisfy the equation.

Within the box the potential is zero, so

$$-\frac{\hbar^2}{2m}\frac{d^2\Psi}{dx^2} = E\Psi(x)$$

which has solutions inside the box of the form,

$$\Psi(x) = A\,e^{ik_1 x} + B\,e^{-ik_1 x}$$

and it is easy to see, by substitution of this solution in the wave equation, that $k_1^2 = 2mE/\hbar^2$.

Outside the box the wave function is,

$$-\frac{\hbar^2}{2m}\frac{d^2\Psi(x)}{dx^2} + V_0\Psi(x) = E\Psi(x)$$

which has solutions of the form,

$$\Psi(x) = C\,e^{k_2 x} + D\,e^{-k_2 x}$$

where

$$k_2^2 = \frac{2m}{\hbar^2}(E - V_0).$$

Now we use the boundary conditions to determine the coefficients A, B, C, D. Outside the box if we consider what happens when $x \to \infty$ we can look at the wave function at infinity. We know that the probability of observing the electron must remain finite so we must have

$$\lim_{x \to \infty} \Psi(x) = 0.$$

So, outside the box we must have,

$$\Psi(x) = D\,e^{-k_2 x} \quad \text{for } x > a$$
$$\Psi(x) = C\,e^{k_2 x} \quad \text{for } x < -a$$

these are functions which decay exponentially with x in both directions moving away from $x = 0$.

The potential well is centred at $x = 0$, so we must have a symmetric distribution of the electrons with respect to x. The observed properties of electrons are really represented by the square of the wave function $|\Psi(x)|^2$. Therefore the wave function squared should be invariant with inversion of x. This constraint merely requires the wave function $\Psi(x)$ itself to be symmetric or antisymmetric with respect to inversion of x.

We will continue by considering only the symmetric (or even parity) wave function solutions (a similar approach can be used for the antisymmetric wave function). This imposes the following condition,

$$\Psi(x) = \Psi(-x)$$
$$A\,e^{ik_1x} + B\,e^{-ik_1x} = A\,e^{-ik_1x} + B\,e^{ik_1x}$$

which implies that $A = B$. Therefore

$$\Psi(x) = A(e^{ik_1x} + e^{-ik_1x})$$
$$\Psi(x) = 2A\cos(k_1x).$$

We then need to apply the boundary conditions at $x = \pm a$. The boundary conditions merely require that Ψ and $d\Psi/dx$ remain continuous throughout.

Therefore at $x = a$ we must have,

$$2A\cos(k_1a) = D\,e^{-k_2a}$$

and

$$-2k_1A\sin(k_1a) = -k_2D\,e^{-k_2a}.$$

These two conditions can then be used to obtain the coefficients A and D. A similar approach can be used at $x = -a$ to obtain a relation between the coefficients A and C. Dividing these equations gives,

$$k_1\tan(k_1a) = k_2$$

and so substituting for k_1 and k_2 from the earlier results gives

$$\frac{2mE}{\hbar^2}\tan^2\left(\frac{(2mE)^{1/2}}{\hbar}a\right) = \frac{2mE}{\hbar^2}(E - V_0)$$

$$E - \tan^2\left(\frac{(2mE)^{1/2}}{\hbar}a\right) = V_0.$$

The result is that for $E > V_0$ the electrons are free with all possible values of energy E although the wave function is perturbed in the vicinity of the potential well. For $E < V_0$ the electrons are contained within the box (with some quantum tunnelling at the boundaries) and have a set of discrete allowed energies determined by the box parameters a and V_0. The situation is perhaps best depicted as shown in Fig. 4.5.

If there is a periodic potential within the square well, then the situation can be described as a perturbation of the above solution. In fact, this case quickly begins to resemble a real solid, with a large square-well barrier at the boundary of the solid and localized potential wells representing the atomic cores. In this case, which is depicted in Figs 5.1 and 5.2, for energies less than V_1 the electrons are highly localized at the 'atomic cores'. For energies between V_1 and V_2 the electrons are quasi-free conduction electrons because they can migrate throughout the 'solid' but are constrained by the solid boundaries. For energies greater than V_2 the electrons are free and can have a continuous range of energies, but these allowed energy states correspond to electrons which have completely escaped from the solid.

Only certain energies are allowed for the 'quasi-free' electrons in the Sommerfeld model because the boundary conditions imposed by the potential well can only be satisfied by certain values of the wave vector k. Since the energy is determined by the wave vector (to a first approximation by $E = \hbar^2 k^2 / 2m$), it follows that only restricted energies of the wave function can meet the boundary conditions.

Exercise 4.3   Electronic specific heat of copper at 300 K. This can be calculated by finding the kinetic energy of the electrons according to the quantum mechanical free-electron model and then differentiating with respect to temperature.

The kinetic energy of the electrons is given classically by,

$$E_K = \tfrac{3}{2} k_B T N$$

where N is the number of electrons. According to quantum mechanics only those electrons close to the Fermi level can contribute to the specific heat, and therefore the value of N that should be used for specific heat calculations is smaller than the total number of electrons. One way of determining the effective number of electrons N^* is to use the density of states close to the Fermi level and make an estimation,

$$N^* = N(E_F) k_B T$$

so that if we suppose that $N(E)$ does not vary too rapidly with energy close to E_F and we include all electrons within an energy range $k_B T$ of the Fermi level, we can find the effective number N^* of electrons contributing to the specific heat capacity. So the collective kinetic energy of these electrons is,

$$E_K = N^* \tfrac{3}{2} k_B T$$
$$E_K = \tfrac{3}{2} k_B^2 T^2 N(E_F)$$

and if kinetic energy is the only relevant energy of the electrons in this model, then the internal energy $U = E_K$, and the electronic heat capacity is therefore

$$C_v^e = \frac{dE_K}{dT} = 3k_B^2 T N(E_F).$$

Here the density of states at the Fermi level is unaffected by temperature but we still need an expression for $N(E_F)$. The expression for $N(E_F)$ that can be used is

$$N(E_F) = \frac{3N}{2E_F}$$

where N is the total number of conduction electrons.* Therefore,

$$C_v^e = \frac{9Nk_B^2 T}{2E_F}$$

and if we consider the molar specific heat then $N = N_0$ the number of conduction electrons per mole. In a monovalent metal such as copper this number is equal to Avogadro's number. We know from Exercise 4.1 that

$$E_F = (3\pi^2 N_{tot})^{2/3} \frac{\hbar^2}{2m}$$

where N_{tot} is the total number of conduction electrons per unit volume in copper. This is $8.5 \times 10^{28}\,\mathrm{m}^{-3}$, and hence $E_F = 1.13 \times 10^{-18}\,\mathrm{J}\,(= 7.05\,\mathrm{eV})$. Inserting the values $k_B = 1.38 \times 10^{-23}\,\mathrm{J\,K}^{-1}$, $T = 300\,\mathrm{K}$, $N_0 = 6.02 \times 10^{23}$ electrons/mole and $E_F = 1.13 \times 10^{-18}\,\mathrm{J}$ into the previous equation we obtain the following molar heat capacity for the electrons

$$C_v^e = 0.151\,\mathrm{J\,mol}^{-1}\,\mathrm{K}^{-1}.$$

If we consider the lattice specific heat C_v^l at low temperatures to be given by the approximate expression obtained from the Debye model,

$$C_v^l = \frac{12\pi^4}{5} k_B N_0 \left(\frac{T}{\theta_D}\right)^3,$$

then the temperature at which the heat capacities of the electrons and lattice are identical is,

$$C_v^l = C_v^e$$

$$\frac{12\pi^4}{5} k_B N_0 \left(\frac{T}{\theta_D}\right)^3 = \frac{9k_B N_0 T}{2T_F}$$

and substituting the given value of the Debye temperature $\theta_0 = 348\,\mathrm{K}$

$$T^2 = 9.942$$

$$T \simeq 3.15\,\mathrm{K}.$$

A second 'mathematical solution' exists at high temperatures when

$$\frac{9}{2} N_0 k_B \frac{T}{T_F} = 3N_0 k_B$$

where T_F is the Fermi temperature $(T_F = E_F/k_B = 81.9 \times 10^3\,\mathrm{K})$ and hence

$$T = \tfrac{6}{9} T_F$$

$$= 0.667 T_F$$

$$T = 54.6 \times 10^3\,\mathrm{K}$$

---

*You can also use the alternative expression $(\pi^2/2)(Nk_B^2)(T/E_F)$ for the electronic heat capacity with only minor differences to the final numerical values.

but the material will be gaseous at this temperature, so it is not a physical solution.

Exercise 5.1 Effective mass of electrons in bands. A free electron with wave vector k, energy E and mass m, will, since it is free and subject to no potential energy, obey the equation,

$$\frac{-\hbar^2}{2m}\nabla^2\psi(x) = E\psi(x)$$

and for a plane wave, free electron, with a wave function,

$$\psi(x) = A\,e^{ikx} + B\,e^{-ikx}$$

this yields the solution,

$$E = \frac{\hbar^2 k^2}{2m}$$

which is a useful relationship between the energy E, wave vector k and mass m.

When an electron is in a solid, however, its movement is affected not only by any external electric fields but also by an internal potential $V(x)$ caused by the periodic atomic array and the other electrons. Therefore, the energy equation becomes more general

$$\frac{-\hbar^2}{2m}\nabla^2\psi(x) + V(x)\psi(x) = E\psi(x).$$

Under these circumstances the relationship between E, k, and m given above no longer holds.

However, there will still be some relation between E and k, and if we compare this with the relationship which holds for free electrons, then we can introduce a correction to this equation by describing the motion in terms of a free electron with a modified effective mass

$$E = \frac{\hbar^2 k^2}{2m^*}$$

where m^* is the effective mass. This happens to be a useful practical result, but we should not lose sight of the fact that it is an artificial procedure in many respects, because it is not really the mass of the electron which is changing but the relationship between E and k, and we are simply choosing to represent this mathematically by incorporating the changes into the expression through the mass of the electron.

If we now consider how the effective masses of the electrons in an energy band are related to the form of the energy band itself, we can obtain another simple relationship. We know that force equals mass times acceleration,

$$F = m^* \frac{dv}{dt}$$

and the velocity can be represented as energy differentiated with respect to momentum,

$$v = \frac{1}{\hbar} \frac{dE}{dk}$$

therefore

$$F = m^* \frac{d}{dt} \left(\frac{1}{\hbar} \frac{dE}{dk} \right)$$

$$= \frac{m^*}{\hbar} \frac{dk}{dt} \frac{d^2E}{dk^2}$$

But, we also know that force is the derivative of momentum with respect to time,

$$F = \frac{dp}{dt} = \frac{d}{dt}(\hbar k)$$

$$F = \hbar \frac{dk}{dt}.$$

Equating the expressions for F gives

$$m^* = \frac{\hbar^2}{\left(\dfrac{d^2E}{dk^2} \right)}.$$

Here (d^2E/dk^2) is the curvature of the energy bands in k-space. Therefore, the effective mass of the electrons is determined solely by the curvature of the electron bands, an interesting and rather surprisingly simple result.

Exercise 5.2 Origin of the electron bands in materials. The description of the emergence of electron bands in solids can be approached either as a perturbation of the free-electron model or as a perturbation of the isolated energy levels of bound electrons within a single atom.

As we have shown in Chapter 5 a free electron can have any energy value and still satisfy the Schrödinger (total energy) wave equation. However, once constraints are put on the electrons, in the form of arbitrary potentials, the allowed energy states become discretized as a requirement for meeting the boundary conditions. In the infinite square-well potential the energy levels closely resemble those in the isolated atom because the electron is constrained by the potential to a limited region of space. When the potential is finite the separation of the energy levels depends entirely on how far below the top of the energy well the electron is. Those deep-lying, low-energy, states are widely separated. Those that are nearer to the top of the well are closer together in energy. The free-electron perturbation works very well for the 'quasi-free' conduction band electrons in a solid, which are those higher energy electrons which migrate throughout the solid.

Alternatively if we begin from the energy levels in a single isolated atom, then these levels are completely separated, as in the infinite potential well. However, as we bring together more and more atoms to form a solid, the potential wells of these atoms start to overlap producing a perturbation of the potential well of the individual atom. This perturbation is more significant for the outer electrons which are at higher energies, but is less so for the inner core electrons.

The result is that for the outer electrons there begins to be some spatial overlap of the electron wave functions, and there is a resulting coalescence of available states which then form an allowed energy band. If there are N identical atoms in the solid, then for each atomic energy level there will arise N energy levels in the solid. By Pauli's exclusion principle no two electrons can have an identical set of quantum numbers and therefore the energies of these N levels will be different. The result is an energy band, which is a quasi-continuous range of allowable energy states. This is shown in Fig. 5.5.

The key to the calculation of the band gaps is that they are equal to the Fourier coefficients of the crystal potential. Therefore for the first band gap we need to find the first coefficient of the Fourier series expansion. For a square wave of periodicity a and with width of $0.8\,a$ and height V, the Fourier expansion is

$$f(x) = A_0 + \sum_{n=1}^{\infty} A_n \cos\left(\frac{n\pi x}{a}\right).$$

The Fourier expansion for a periodic square-well potential can be found in 'CRC Standard Mathematical Tables' by W. H. Beyer, 27th edition, 1984 on page 403. The expansion is

$$f(x) = -V\left\{0.8 + \frac{2}{\pi}\sum_{n=1}^{\infty}\frac{(-1)^n}{n}\sin(0.8\,n\pi)\cos\left(\frac{n\pi x}{a}\right)\right\}.$$

So the first Fourier coefficient, with $n = 1$, is,

$$E_g = V\frac{2}{\pi}\sin(0.8\,\pi)$$

$$= 0.374\,V$$

and since $V = 2$ electron volts the energy gap is

$$E_g = 0.748\,\text{eV}.$$

Exercise 5.3 Number of conduction electrons in a Fermi sphere of known radius. Since the problem states that the metal is quasi-free electron metal, and has a spherical Fermi surface in the space, then the free-electron approximation can be used to give the density of states in a volume V of a material. As given in section 4.4.7 this is

$$\frac{dN_0(E)}{dE} = D(E) = \frac{V}{4\pi^2}\left(\frac{2m}{\hbar^2}\right)^{3/2} E^{1/2}$$

where V is volume of the sample. In this case we will use only a unit cell of the simple cubic lattice, therefore, if the lattice parameter is a the volume will be

$$V = a^3.$$

Integrating the above equation and setting E equal to the Fermi energy E_F, this gives the total number of energy states

$$N_0(E_F) = \frac{V}{6\pi^2}\left(\frac{2m}{\hbar^2}\right)^{3/2} E_F^{3/2}.$$

Allowing two electrons per state, one with spin up the other with spin down, the total number of conduction electrons will be $N = 2N_0(E_F)$

$$N = \frac{V}{3\pi^2}\left(\frac{2m}{\hbar^2}\right)^{3/2} E_F^{3/2}.$$

At this stage we note that for free electrons $E = \hbar^2 k^2/2m$, so that the Fermi energy is

$$E_F = \frac{\hbar^2 k_F^2}{2m}.$$

Consequently the total number of conduction electrons is

$$N = \frac{V}{3\pi^2} k_F^3.$$

At the Brillouin zone boundary of this simple cubic lattice we must have

$$k_F = \frac{\pi}{a}.$$

Therefore substituting in the values for V and k_F we arrive at the following number of conduction electrons per atom needed to just cause the Fermi surface to touch the Brillouin zone boundary

$$N = \frac{\pi}{3} = 1.047.$$

Exercise 6.1 Brillouin zones in a two-dimensional lattice. If the lattice parameters in real space are $a = 0.2 \times 10^{-9}$ m and $b = 0.4 \times 10^{-9}$ m, then the reciprocal lattice vectors are

$$k_x = \frac{\pi}{a}$$

$$k_y = \frac{\pi}{b}.$$

The reciprocal lattice of a rectangular lattice is also rectangular.

Dimensions of the first Brillouin zone are twice the lengths of the reciprocal lattice vectors

$$k_x = \frac{2\pi}{a} = 31.4\,\text{nm}^{-1}$$

$$k_y = \frac{2\pi}{b} = 15.7\,\text{nm}^{-1}.$$

If the atom has a valence of 1 then it will have one electron per atom in the conduction band. This conduction band is as defined in the statement of the problem a free-electron sphere. In two dimensions the free electrons will occupy a circle, the radius of this 'Fermi circle' is k_F. Therefore

'area' of Fermi circle $= \pi k_F^2$.

The total number of available k states contained in a free-electron circle radius k_F is simply the area of the Fermi circle πk_F^2 divided by the area of a k-state. The area of each k-state is

$$A_k = \left(\frac{2\pi}{a}\right)\left(\frac{2\pi}{b}\right) = \frac{4\pi^2}{ab}.$$

Therefore the number of k-states between zero energy and the Fermi energy E_F in this two-dimensional case is

$$N_0(E_F) = \frac{ab}{4\pi} k_F^2.$$

The total number of electrons will be double this number because of the possibility of accommodating a 'spin-up' and 'spin-down' electron at each k-state

$$N_0(E_F) = \frac{ab}{2\pi} k_F^2$$

and since the number of electrons must be 1, $N_0(E_F) = 1$. Therefore

$$k_F = \sqrt{\frac{2\pi}{ab}}$$

$$k_F = \sqrt{\frac{2\pi}{0.08 \times 10^{-18}}}$$

$$k_F = 8.86 \, \text{nm}^{-1}.$$

Since the dimensions of the first Brillouin zone are $k_x = 15.7 \, \text{nm}^{-1}$ and $k_y = 7.85 \, \text{nm}^{-1}$, the free-electron sphere crosses the zone boundary along the k_y direction, but does not cross the zone boundary along the k_x direction. This is shown in Fig. S1. The electron band structure is therefore as shown in Fig. S2.

Exercise 6.2   Number of k states in reciprocal space. Consider a simple cubic lattice with lattice parameter a and with N^3 primitive cells. This will give a cube of side L such that

$$L = Na.$$

If we apply periodic boundary conditions the allowed values of k are as follows:

$$k_x, k_y, k_z = 0, \pm \frac{2\pi}{L}, \pm \frac{4\pi}{L}, \pm \cdots \pm \frac{2N\pi}{L}.$$

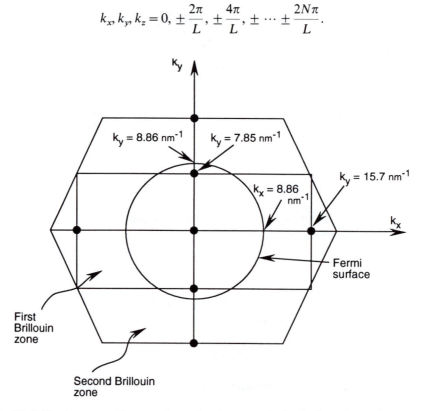

Fig. S1 Brillouin zones of the two-dimensional rectangular lattice in reciprocal space, and the Fermi 'sphere' superimposed on the zones.

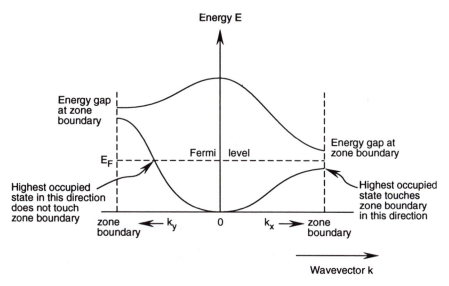

Fig. S2 Electron band structure of the two-dimensional rectangular lattice.

Ignoring the case $k = 0$, which is physically identical to $k = 2N\pi/L$, this gives N allowed k-states for each of k_x, k_y, k_z. That is there are N^3 allowed k-states. Since all of the allowed states can be represented in the first Brillouin zone in the reduced zone scheme this means that the number of k-states in the first Brillouin zone of the reduced zone scheme is N^3.

Alternatively we may say that the volume of k-space up to the Nth Brillouin zone is

$$V_N = \left(\frac{2\pi N}{a}\right)^3.$$

Futhermore, it is known that the volume occupied by a k-state in reciprocal space is

$$V_k = \left(\frac{2\pi}{a}\right)^3.$$

Therefore the number of allowed k-states is simply the volume of the k-space representation of the crystal up to the Nth Brillouin zone divided by the volume occupied by a single k-state:

$$\text{No. of } k\text{-states} = N^3.$$

Exercise 6.3 Fermi energy of sodium and aluminium. If we assume that both metals are free-electron-like, then in both cases the Fermi energy is related to the wave vector k_F at the Fermi surface by the equation

$$E_F = \frac{\hbar^2}{2m} k_F^2.$$

The total number of k-states contained in a volume of k-space of radius k_F is equal to the volume of the Fermi sphere in k-space ($\frac{4}{3}\pi k_F^3$) divided by the volume of a Brillouin zone $(8\pi^3/a^3) = (8\pi^3/V)$

$$N_0(k_F) = \left(\frac{4\pi}{3} k_F^3\right) \bigg/ (8\pi^3/a^3)$$

$$N_0(k_F) = \frac{a^3}{6\pi^2} k_F^3$$

where V is the volume of the specimen and a is the linear dimension of a unit cell. Since each electron can have 'spin up' or 'spin down', the number of electrons that can be contained in such a volume is twice the number of k-states

$$N_0(k_F) = \frac{V}{3\pi^2} k_F^3$$

If we consider a unit cell of each metal, then the volume of this crystallographic unit cell is $V = (0.43 \times 10^{-9})^3 = 79.5 \times 10^{-30}\,\text{m}^3$ for sodium, and $V = (0.4 \times 10^{-9})^3 = 64 \times 10^{-30}\,\text{m}^3$ for aluminium. Within these unit cells there are effectively two atoms in a bcc lattice and four atoms in an fcc lattice. Therefore, the above unit cells contain 2 conduction electrons in sodium but 12 conduction electrons in aluminium because aluminium is trivalent.

For the case of sodium therefore we must have $N_0(k_F) = 2$ and as,

$$k_F^3 = \frac{6\pi^2}{79.5 \times 10^{-30}}\,\text{m}^{-3}$$

therefore,

$$k_F = 9.06 \times 10^9\,\text{m}^{-1}$$

and consequently the Fermi energy is

$$E_F = \hbar^2 k_F^2 / 2m$$

$$E_F = \frac{(1.054 \times 10^{-34})^2}{2(9.109 \times 10^{-31})} (9.07 \times 10^9)^2\,\text{joules}$$

$$= 5.016 \times 10^{-19}\,\text{joules}$$

$$E_F = 3.13\,\text{eV}.$$

For the case of aluminium we must have $N_0(k_F) = 12$ and so,

$$k_F^3 = \frac{36\pi^2}{64 \times 10^{-30}}$$

therefore,

$$k_F = 17.71 \times 10^9 \, m^{-1}$$

and the Fermi energy is

$$E_F = \frac{(1.054 \times 10^{-34})^2}{2(9.109 \times 10^{-31})}(17.71 \times 10^9)^2$$

$$= 1.913 \times 10^{-18} \, joules$$

$$E_F = 11.9 \, eV.$$

Exercise 7.1 Approximation to the Fermi function in semiconductors. The Fermi function $f(E)$, which describes the probability that an electron occupies an energy level E at temperature T is

$$f(E) = \frac{1}{1 + \exp\left(\dfrac{E - E_F}{k_B T}\right)}$$

where E_F is the Fermi level. If we consider this simply in terms of a probability as a function of ΔE then

$$f(\Delta E) = \frac{1}{1 + \exp\left(\dfrac{\Delta E}{k_B T}\right)}.$$

In the example under consideration the band gap is $0.5 \, eV$. Considering the lowest available energy state in the conduction band and locating the Fermi level at the mid-point of the band gap, it is clear that the lowest possible value of ΔE for states in the conduction band is $0.25 \, eV$ above the Fermi level

$$\Delta E \geqslant 0.25 \, eV.$$

Consequently for all energy levels in the conduction band $(\Delta E / k_B T) \geqslant 9.67$ and so $\exp(\Delta E / k_B T) \geqslant 15.8 \times 10^3$. Therefore the exponential term dominates the constant term in the denominator and,

$$f(\Delta E) = \frac{1}{1 + \exp\left(\dfrac{\Delta E}{k_B T}\right)} \approx \exp\left(\frac{-\Delta E}{k_B T}\right).$$

Since this is true for the energy levels just above the band gap it will also be true for higher energy levels, and for temperatures of $300 \, K$ and below. Note however that for electrons at the bottom of the conduction band $\Delta E = E_g/2$ and not $\Delta E = E_g$ so that the probability of occupancy of energy levels at the bottom of the conduction band can also be written

$$f(E_g) = \exp\left(\frac{-E_g}{2k_B T}\right).$$

Exercise 7.2   Temperature dependence of conductivity in intrinsic semiconductors.
If the wavelength of the absorption edge is $\lambda = 1771\,\text{nm}$, then the band gap energy will be

$$E_g = \frac{hc}{\lambda}$$

$$= \frac{(2.99 \times 10^8)(6.62 \times 10^{-34})}{1.771 \times 10^{-6}}$$

$$= 1.12 \times 10^{-19}\,\text{joules}$$

$$E_g = 0.698\,\text{eV}.$$

Since the material is described as an intrinsic conductor, its conductivity will obey the relation

$$\sigma = \sigma_0 \exp\left(\frac{-E_g}{2k_B T}\right).$$

If we let the conductivity increase by 30% at a temperature $T + \Delta T$, then,

$$1.3 = \frac{\exp\left(-E_g/2k_B(T+\Delta T)\right)}{\exp\left(-E_g/2k_B T\right)}$$

$$1.3 = \exp\left(\frac{-E_g}{2k_B}\left(\frac{1}{T+\Delta T} - \frac{1}{T}\right)\right)$$

therefore,

$$T + \Delta T = \frac{1}{\left(\dfrac{1}{T} - \dfrac{2k_B}{E_g}\log_e(1.3)\right)}$$

$$T + \Delta T = 305.9\,\text{K}$$

consequently a temperature rise of $\Delta T = 5.9\,\text{K}$ from 300 K to 305.9 K will lead to a 30% rise in conductivity.

Exercise 7.3   Electronic properties of gallium arsenide, silicon and germanium.
Table S1 shows a comparison of the electronic properties of the three materials.

Applications in which gallium arsenide has an advantage over the others include those involving emission and absorption of light (e.g. lasers and optical communications), in which the direct band gap gives a higher probability of transition, and those requiring fast response times, in which the high electron mobilities in GaAs are advantageous (e.g. high-speed computer applications).

Table S1

Property	Si	Ge	GaAs
Band gap (eV)	1.1	0.7	1.4
Electron mobility (m²/Vs)	0.15	0.39	0.85
Effective mass of electrons m^*/m_e	0.97	1.64	0.07
Effective mass of holes m^*/m_h	0.6	0.3	0.5
k at valence band maximum	0	0	0
k at conduction band minimum	0	at zone edge in (1, 1, 1)	near zone edge in (1, 0, 0)
Electrical conductivity at 300 K ($\Omega^{-1}\,m^{-1}$)	9×10^{-4}	2.2	1×10^{-6}
Absorption edge (nm)	1104 (IR)	1873 (IR)	871 (visible)

Since the relation between the electron energy E and the wave vector k is given as

$$E = Ak^2$$

we can calculate the effective mass from the equation,

$$m^* = \frac{\hbar^2}{(d^2 E/dk^2)}$$

where,

$$\frac{d^2 E}{dk^2} = 2A$$

therefore,

$$m^* = \frac{\hbar^2}{2A}.$$

With $A = 7.5 \times 10^{-38}\,\mathrm{J\,m^2}$,

$$m^* = 74 \times 10^{-33}\,\mathrm{kg}$$

and since the rest mass of an electrons is,

$$m_0 = 9.1 \times 10^{-31}\,\mathrm{kg}$$

we have

$$\frac{m^*}{m_0} = 0.08.$$

Exercise 8.1 Drift velocity of conduction electrons. If we use the free-electron approximation, then the following relation can be used,

$$\frac{1}{\rho} = \sigma = \frac{Ne^2}{m} \tau_F$$

where ρ is resistivity, σ is conductivity, N is the number of conduction electrons per unit volume, e is the electronic charge, m the electronic mass and τ_F is the mean time between collisions at the Fermi level.

If we let l be the mean free path of the conduction electrons then,

$$v_F = \frac{l}{\tau_F}$$

where v_F is the Fermi velocity. Therefore

$$\rho = \frac{mv_F}{Ne^2 l}.$$

The energy and velocity of electrons at the Fermi level are related by the equation,

$$E_F = \frac{1}{2} mv_F^2$$

so,

$$v_F = \sqrt{2E_F/m}$$

and substituting $E_F = 1.922 \times 10^{-18}\,\text{J}\,(= 12\,\text{eV})$ and $m = 9.1 \times 10^{-31}\,\text{kg}$,

$$v_F = 2.05 \times 10^6\,\text{m}\,\text{s}^{-1}$$

and rearranging the equation for the resistivity

$$l = \frac{mv_F}{Ne^2 \rho}.$$

This means that we need to calculate N, the number of conduction electrons per unit volume. Since each atom gives three conduction electrons to the conduction band,

$$N = 3N_a$$

where N_a is the number density of atoms. The density of $\text{Al} = 2700\,\text{kg}\,\text{m}^{-3}$. Therefore 6.02×10^{26} atoms weigh $27\,\text{kg}$. Hence the number density of aluminium atoms is,

$$N_a = 6.02 \times 10^{28}\,\text{atoms per m}^3$$

and the number of conduction electrons per unit volume N is three times greater, because aluminium is trivalent:

$$N = 3N_a$$

$$N = 1.806 \times 10^{29}\,\text{m}^{-3}.$$

The mean free path length is then,

$$l = \frac{mv_F}{Ne^2\rho}$$

$$l = \frac{(9.1 \times 10^{-31})(2.05 \times 10^6)}{(1.806 \times 10^{29})(1.602 \times 10^{-19})^2(3 \times 10^{-8})}$$

$$l = 13.4 \times 10^{-9}\,\text{m}.$$

The mean free path of the electrons is related to the mean drift velocity v_d under the action of an electric field ξ. The current density J is given by,

$$J = \sigma\xi$$

where σ is the conductivity. Furthermore

$$J = Nev_d$$

where N = number density of conduction electrons, e = electronic charge and v_d = drift velocity.

$$v_d = \frac{\sigma\xi}{Ne} = \frac{\xi}{\rho Ne}$$

$$= \frac{(1 \times 10^3)}{(3 \times 10^{-8})(18 \times 10^{28})(1.6 \times 10^{-19})}$$

$$= 1.15\,\text{m s}^{-1}.$$

Exercise 8.2   Conductivity in intrinsic and extrinsic semiconductors. The problem states that the doped semiconductor, the n-type germanium, contains 10^{23} ionized donors per cubic metre. This means that it has 10^{23} electrons per cubic metre in the conduction band. The intrinsic semiconductor will have a number n of electrons per unit volume in its conduction band, where n is determined by the temperature and the electronic band gap.

The conductivity of the extrinsic germanium σ_{ex} is determined only by the electrons according to the relation

$$\sigma_{ex} = Ne\mu_e$$

where N is the number of electrons per unit volume in the conduction band, e is the electronic charge and μ is the electron mobility. In this case $N = 1 \times 10^{23}\,\text{m}^{-3}$, $e = 1.602 \times 10^{-19}\,\text{C}$ and $\mu_e \simeq 0.39\,\text{m}^2\,\text{s}^{-1}\,\text{V}^{-1}$ for electrons, and $\mu_h = 0.19\,\text{m}^2\,\text{s}^{-1}\,\text{V}^{-1}$ for holes.

In the intrinsic germanium the conductivity is determined by both the electrons and holes, and is given by

$$\sigma_{in} = eN_e\mu_e + eN_h\mu_h$$

where N_e is the number density of electrons in the conduction band and μ_e is their mobility, and N_h is the number density of holes in the valence band and μ_h is their mobility. It can reasonably be assumed that $N_e = N_h = N$ in an intrinsic semiconductor, and also that e is the same for electrons and holes

$$\sigma_{in} = Ne(\mu_e + \mu_h).$$

The number density of electrons in the conduction band N can be obtained from thermodynamic considerations. Using Fermi–Dirac statistics the number density is then given by the equation,

$$N(E) = 2f(E)D(E)$$

$$= \frac{2}{1 + \exp((E - E_F)/k_B T)} \cdot \frac{\pi}{4} \cdot \left(\frac{2m}{\pi^2 \hbar^2}\right)^{3/2} E^{1/2}$$

and assuming that the exponential term in the denominator is much greater than unity,

$$N(E) \simeq \frac{1}{2\pi^2} \left(\frac{2m}{\hbar^2}\right)^{3/2} E^{1/2} \exp\left(\frac{-(E - E_F)}{k_B T}\right).$$

Integrating this equation, assuming that $E = k_B T$ and $E - E_F \simeq E_g/2$ for the conduction electrons at the bottom of the conduction band, leads to

$$N = \frac{1}{4} \left(\frac{2m}{\hbar^2}\right)^{3/2} (k_B T)^{3/2} \left(\frac{1}{\pi}\right)^{3/2} \exp\left(\frac{-E_g}{2k_B T}\right).$$

Then rearranging and correcting for the effective mass of the electrons gives

$$N = \frac{1}{4} \left(\frac{2mk_B}{\pi\hbar^2}\right)^{3/2} \left(\frac{m^*}{m}\right)^{3/2} T^{3/2} \exp\left(\frac{-E_g}{2k_B T}\right).$$

The first term on the right-hand side is a constant with value 4.82×10^{21} $K^{-3/2} m^{-3}$ so this gives the final expression for N as

$$N = 4.82 \times 10^{21} \left(\frac{m^*}{m_0}\right)^{3/2} T^{3/2} \exp\left(\frac{-E_g}{2k_B T}\right)$$

where m^* is the effective mass of electrons and m_0 is the rest mass of free electrons. So the conductivity becomes,

$$\sigma_{in} = 4.82 \times 10^{21} \left(\frac{m^*}{m_0}\right)^{3/2} T^{3/2} e(\mu_e + \mu_h) \exp\left(\frac{-E_g}{2k_B T}\right)$$

and $m^*/m_0 \approx 0.8$, $T = 300\,K$, $\mu_e = 0.36\,m^2\,s^{-1}\,V^{-1}$, $\mu_h = 0.18\,m^2\,s^{-1}\,V^{-1}$, $E_g = 0.7\,eV$ (1.12×10^{-19} joules), $e = 1.602 \times 10^{-19}\,C$. Therefore,

$$\sigma_{in} = (4.82 \times 10^{21})(0.71)(5.196)(1.602 \times 10^{-19})(0.58) \exp(-13.53)$$

$$\sigma_{in} = 2.2\,\Omega^{-1}\,m^{-1}$$

and

$$\sigma_{ex} = (1 \times 10^{23})(1.602 \times 10^{-19})(0.39)$$

$$= 6.248 \times 10^3 \, \Omega^{-1} \, m^{-1}.$$

Therefore the ratio of extrinsic to intrinsic conductivities is

$$\sigma_{ex}/\sigma_{in} = 0.3 \times 10^4.$$

Exercise 8.3   Thermoluminescence and lifetime of electrons in traps. The lifetime of electrons is related to the depth of traps below the conduction band is ΔE, such that,

$$\tau(T) = \frac{1}{s} \exp\left(\frac{\Delta E}{k_B T}\right)$$

therefore

$$\Delta E = k_B T \log_e (s\tau)$$

$$= (1.38 \times 10^{-23})(273) \log_e ((4.64 \times 10^{17})(1.0 \times 10^{10}))$$

$$\Delta E = (1.38 \times 10^{-23})(273)(63.70)$$

$$\Delta E = 2.4 \times 10^{-19} \, \text{joules}$$

$$\Delta E = 1.5 \, \text{eV}.$$

So the depth of traps below the conduction band is 1.5 eV (or 2.4×10^{-19} joules).

Using the empirical relationship of Urbach, the temperature at which the peak of the glow curve occurs, T^*, is

$$T^* = 500 \, \Delta E$$

where ΔE is measured in electron volts and T^* is in Kelvins. Therefore

$$T^* = 750 \, \text{K}$$

or 477 °C.

The lifetime $\tau(T)$ of electrons in traps at a given temperature T is

$$\tau(T) = \frac{1}{s} \exp\left(\frac{\Delta E}{k_B T}\right).$$

Therefore, if the lifetime at 273 K and 373 K are compared,

$$\frac{\tau(373)}{\tau(273)} = \frac{\exp(\Delta E/(373 k_B))}{\exp(\Delta E/(273 k_B))}$$

$$= \exp\left(\frac{\Delta E}{k_B} \cdot \left(\frac{1}{373} - \frac{1}{273}\right)\right)$$

$$\tau(373) = (1.0 \times 10^{10}) \exp\left(\frac{2.4 \times 10^{-19}}{1.38 \times 10^{-23}}(0.00268 - 0.00366)\right)$$

$$= (1.0 \times 10^{10}) \exp(-17.04)$$

$$\tau(373) = 3.96 \times 10^2$$

$$= 396 \text{ seconds.}$$

Example 9.1　Optical properties of metals and insulators. The various optical constants $\alpha, \delta, R, \varepsilon_1$ and ε_2 can be determined from the refractive index n and the extinction coefficient k by the following equations

$$\alpha = \frac{4\pi k}{\lambda} \text{ (m)}$$

$$\delta = \frac{\lambda}{4\pi k} \text{ (m}^{-1})$$

$$R = \frac{(n-1)^2 + k^2}{(n+1)^2 + k^2} \text{ (dimensionless)}$$

$$\varepsilon_1 = n^2 - k^2 \text{ (dimensionless)}$$

$$\varepsilon_2 = 2nk \text{ (dimensionless).}$$

The values for the four materials, as calculated from n and k at $\lambda = 1240$ nm ($\hbar\omega = 1$ eV), are shown in Table S2.

Materials 1 and 2 have very high reflectance R and high absorption ε_2 at 1 eV and are therefore metals. Both materials 3 and 4 have low reflectance R and absorption ε_2, and a very low extinction coefficient k. These therefore must have a band gap greater that 1 eV which prevents absorption of light at this wavelength. Both are therefore insulators.

In fact material 1 is aluminium, 2 is gold and 3 and 4 are different types of glass.

Table S2

Material	n	k	R	α
1	1.21	12.46	0.94	0.126×10^9
2	0.13	8.03	0.99	0.08×10^9
3	1.51	1.12×10^{-6}	0.04	11.35
4	1.92	1.5×10^{-6}	0.10	15.21

Material	δ	ε_1	ε_2
1	7.92×10^{-9}	-153.8	30.15
2	12.28×10^{-9}	-64.5	2.09
3	0.088	2.28	3.4×10^{-6}
4	0.066	3.69	5.8×10^{-6}

Exercise 9.2   Classification of principal electronic transitions. The principal types of electronic transitions that can occur are as follows:

(a) Interband transitions
   (i) High-energy transitions (bottom of valence band to top of conduction band)
   (ii) Band gap edge transitions (top of valence band to bottom of conduction band)
(b) Impurity level transitions
   (i) Excition generation (valence band to trap)
   (ii) Impurity level excitation (trap to conduction band)
(c) Intraband transitions (metals only)
   (i) Transitions within partially filled band.

Characteristic colours

The characteristic colours of materials are determined mainly by the band gap energy. Absorption, and hence reflectance, can only occur when there is an allowed electronic transition of the appropriate energy. Therefore in large band gap materials (e.g. diamond) all optical wavelengths are transmitted. As the band gap becomes smaller however, certain wavelengths at the shorter wavelength, higher energy end of the spectrum start to get absorbed. This can lead to green, yellow, orange, and red transmission as the band gap decreases. Finally the transmission in the visible range goes to zero as the band gap falls below the visible red end of the spectrum at about 1.7 eV, leading to a black colour.

Colours by reflection, however, will be different because absorption of a given wavelength allows it to be reflected. Therefore, semiconductors with band gaps in the range 2.5–3.0 eV may appear bluish by reflection but yellow or orange by transmission.

Certain coloured metals such as copper and gold also have their reflectances altered by the presence of characteristic interband transitions at the appropriate energies to cause a yellow or reddish tinge in their reflectance spectrum.

Exercise 9.3   Identification of material from optical absorption spectrum. The optical spectrum is that of an insulator or semiconductor because of the low absorption at long wavelengths (low energies) and higher absorption at short wavelengths (high energies).

The absorption edge, which corresponds to the energy at which electrons can just begin to cross the band gap occurs in this material at about 0.75 eV, where the absorption increases from zero as the photon energy increases. From the data on the band gaps and the absorption edges of the three materials A, B and C, as given in the table, the spectrum must correspond to material B.

In fact the remainder of the data in the tables is irrelevant for interpreting which of the materials corresponds to the given spectrum.

The material will be transparent for wavelengths at which the absorption ε_2

is close to zero. This corresponds to all wavelengths longer than 1653 nm (equivalent to 0.75 eV). As a result it will be opaque, and hence reflecting, for all shorter wavelengths, including the optical region of the spectrum (750–450 nm).

The materials are: A – silicon, B – germanium and C – gallium arsenide.

Exercise 10.1   Strength of exchange field in iron. A relationship exists between the Curie temperature of a ferromagnet and the exchange interaction. If the exchange field H_{ex} is given by

$$H_{ex} = \alpha M$$

and the paramagnetic susceptibility is

$$\chi = \frac{C}{T - T_c} = \frac{M}{H + H_{ex}}$$

where C is the Curie constant, T is the absolute temperature and T_c is the Curie temperature. Therefore

$$C = \frac{N \mu_0 m^2}{3 k_B}$$

where N = number of atoms per unit volume, μ_0 is the permeability of free space, k_B is Boltzmann's constant and m is the moment per atom.

From the Curie–Weiss law equation

$$T_c = \alpha C = \frac{\alpha N \mu_0 m^2}{3 k_B}$$

and therefore

$$\alpha = \frac{3 k_B T_c}{N \mu_0 m^2}.$$

Since $H_{ex} = \alpha M$, and within a single domain the magnetization is saturated so that $M = M_s = N m$, this leads to

$$H_{ex} = \alpha M_s$$

$$= \frac{3 k_B T_c}{\mu_0 m}.$$

The value of m needs to be in A m^2 instead of Bohr magnetons. This can be calculated from the relation 1 Bohr magneton = 9.27×10^{-24} A m^2. Therefore for iron,

$$m = 2.04 \times 10^{-23} \, \text{A m}^2$$

$$H_{ex} = \frac{(3)(1.38 \times 10^{-23})(1043)}{(12.57 \times 10^{-7})(2.04 \times 10^{-23})}$$

Table S3

	Isolated ion magnetic moment (Bohr magnetons)	*Saturation magnetization of bulk material (Am^{-1})*	*Calculated moment per atom in bulk material (Bohr magnetons)*
Fe^{2+}	5.4	1.71×10^6	2.22
Co^{2+}	4.8	1.42×10^6	1.72
Ni^{2+}	3.2	0.48×10^6	0.54

$$= 1.68 \times 10^9 \, A \, m^{-1}.$$

which is a surprisingly high value of magnetic field.

Exercise 10.2   Comparison of the magnetic moments on atoms in bulk form and in isolation. The values of saturation magnetization M_s of the three metals in bulk and the magnetic moments of the isolated atoms of each metal are given in Table S3.

The magnetic moment per atom in the bulk, which is calculated from the saturation magnetization by dividing by the number of atoms per unit volume in the metal is also shown.

The important result here is that the magnetic moments on the atoms in bulk form (e.g. solids) are substantially different from the magnetic moments on the same atoms in isolation. Therefore the interactions between the electrons on atoms in bulk material cause significant modifications of the observed magnetic moments.

If we consider the electronic structure of the iron, cobalt, and nickel specimens, we find that the isolated atoms have the numbers of electrons in the outer shells shown in Table S4.

The magnetic properties of these metals are due to the d electrons. However in bulk material the 3d and 4s electrons occupy similar energy levels, and in fact the broadening of the 3d and 4s levels in the solid causes these levels to overlap. This leads to s–d electron mixing, also known as 'hybridization'.

In the solid therefore the 3d and 4s levels can be treated as a single energy band, with twelve possible electron states. Clearly the difference in the number

Table S4

Element	3d	4s
Fe	6	2
Co	7	2
Ni	8	2

Table S5

Element	$(n_+ - n_-)$	$(n_+ + n_-)$
Fe	2.22	8
Co	1.72	9
Ni	0.54	10

Table S6

Element	*Spin-up*	*Spin-down*
Fe	5.11	2.89
Co	5.36	3.64
Ni	5.27	4.73

of spin-up electrons n_+ and spin-down electrons n_- will give the magnetic moment per atom (Table S5).

The number of spin-up and spin-down electrons in the 3d/4s band is as shown in Table S6.

Exercise 10.3   Spontaneous magnetization and the exchange field. The Langevin expression for the magnetization M of a classical paramagnet with localized magnetic moments m on each atomic site is

$$M = Nm\left\{\coth\left(\frac{\mu_0 m \cdot H}{k_B T}\right) - \left(\frac{k_B T}{\mu_0 m \cdot H}\right)\right\}.$$

If an exchange field H_{ex} is introduced that is proportional to the spontaneous magnetization M within a domain, then the effective field becomes,

$$H_{eff} = H + \alpha M.$$

Substituting this into the Langevin expression gives

$$M = Nm\left\{\coth\left(\frac{\mu_0 m \cdot (H + \alpha M)}{k_B T}\right) - \left(\frac{k_B T}{\mu_0 m \cdot (H + \alpha M)}\right)\right\}.$$

For spontaneous magnetization ordering will occur in the absence of an external field. Setting the external field H equal to zero in this equation gives,

$$M = Nm\left\{\coth\left(\frac{\mu_0 m \alpha M}{k_B T}\right) - \frac{k_B T}{\mu_0 m \alpha M}\right\}$$

$$M = Nm\left\{\coth\left(\frac{\mu_0 \alpha m^2 N}{k_B T}\right) - \frac{k_B T}{\mu_0 \alpha N m^2}\right\}.$$

In order for ferromagnetic ordering to occur, within a domain there must be a spontaneous magnetization. Consequently dM/dH must be infinite at the origin of the M, H plane

$$\left(\frac{dM}{dH}\right)_{\substack{H=0 \\ M=0}} = \frac{Nm}{(3k_B T/\mu_0 m) - \alpha Nm}.$$

Therefore, for ferromagnetism to occur we must have the denominator equal to zero (actually negative values also give ferromagnetism). This means that α must have a value of,

$$\alpha \geqslant \frac{3k_B T}{\mu_0 N m^2} = \frac{(3)(1.38 \times 10^{-23})(300)}{(4\pi \times 10^{-7})(9 \times 10^{28})(2 \times 10^{-23})^2}$$

therefore for ferromagnetic ordering to occur we must have

$$\alpha \geqslant 274.5.$$

Subject Index

Page numbers appearing in **bold** refer to figures and page numbers appearing in *italic* refer to tables.

Author Index